D0505001

Solving Problems in Fluid Mechanics

Volume 2

Solving Problems in Fluid Mechanics

Volume 2

J. F. Douglas MSc PhD DIC ACGI CEng MICE MIMechE MIStructE

Longman Scientific &
Technical

Longman Scientific & Technical
Longman Group UK Limited
Longman House, Burnt Mill, Harlow
Essex CM20 2JE, England
and Associated Companies throughout the world

First published by Pitman Publishing Limited in 1970 under the title
Solution of Problems in Fluid Mechanics Part 2
All metric edition first published i. 375
Ninth impression 1984
This edition published by Longman Scientific & Technical in 1986 under the title
Solving Problems in Fluid Mechanics Volume 2

British Library Cataloguing in Publication Data
Douglas, J.F.
  [Solution of problems in fluid mechanics]
  Solving problems in fluid mechanics. – All-metric
  ed. – (Solving problems)
  Vol. 2
  1. Fluid mechanics
  I. [Solution of problems in fluid mechanics]
  II. Title
  620.1′06   TA357

ISBN 0-582-28643-3

Produced by Longman Singapore Publishers (Pte) Ltd.
Printed in Singapore.

Contents

Contents

Preface

The treatment adopted in this second volume is exactly the same as that employed so successfully in the first volume, the subject matter of each section being presented in the form of question and answer. The reader will find all the definitions and theory required, together with selected problems which are fully worked out, and plenty of exercise questions with numerical answers on which to practice and develop skill and understanding.

The material included in this volume covers more advanced work in Fluid Mechanics for engineering students in Universities, Polytechnics and Colleges of Higher Education. The fullness of the treatment has in some places had to be restricted owing to the limited space available. The reader seeking further information in any particular field will find it helpful to refer to "Fluid Mechanics" by Douglas, Gasiorek and Swaffield (Pitman 2nd. Edn 1985).

I would again like to express my appreciation of the assistance which I have received from my former colleagues in the teaching profession. I am particularly indebted to Dr. R.D. Matthews for his advice on the preparation of this new text and for the provision of examples and exercises with particular reference to Chapter 9.

I hope that my readers will not hesitate to let me know of any difficulties that they may experience with this text and I will be glad to receive any constructive criticism.

John Douglas   September 1985

1

Dimensional analysis

Dimensional analysis is a mathematical method which is of considerable value in problems which occur in fluid mechanics. As explained in Part One all quantities can be expressed in terms of certain primary quantities which in mechanics are Length (L), Mass (M) and Time (T). For example

$$\text{Force} = \text{mass} \times \text{acceleration}$$
$$= \text{mass} \times \text{length/time}^2$$

Thus the dimensions of Force will be MLT^{-2}.

In any equation representing a real physical event every term must contain the same powers of the primary quantities (L, M and T). In other words, like must be compared with like or else the equation is meaningless, although it may balance numerically.

This principle of homogeneity of dimensions can be used, (1) to check whether an equation has been correctly formed, (2) to establish the form of an equation relating a number of variables, and (3) to assist in the analysis of experimental results.

1.1 Checking equations

Show by dimensional analysis that the equation

$$p + \tfrac{1}{2}\rho v^2 + \rho gz = H$$

is a possible relationship between the pressure p, velocity v and height above datum z for frictionless flow along a streamline of a fluid of mass density ρ, and determine the dimensions of the constant H.

Solution. If the equation represents a physically possible relationship each term must have the same dimensions and therefore contain the same powers of the primary quantities L, M and T.

The procedure to be adopted is first to determine the dimensions of each of the variables in terms of L, M and T, and then to examine the dimensions of each term in the equation.

The dimensions of the variables are

$$\text{Pressure } p = \frac{\text{force}}{\text{area}} = \frac{\text{mass} \times \text{acceleration}}{\text{area}}$$
$$= ML^{-1}T^{-2}$$

$$\text{Mass density } \rho = \frac{\text{mass}}{\text{volume}} = ML^{-3}$$

$$\text{Velocity } v = \frac{\text{length}}{\text{time}} = LT^{-1}$$

$$\text{Gravitational acceleration } g = LT^{-2}$$

$$\text{Height above datum } z = L$$

The dimensions of each term on the left-hand side are

$$p = ML^{-1}T^{-2}, \quad \tfrac{1}{2}\rho v^2 = ML^{-3} \times L^2 T^{-2} = ML^{-1}T^{-2}$$
$$\rho g z = ML^{-3} \times LT^{-2} \times L = ML^{-1}T^{-2}$$

Thus all terms have the same dimensions and the equation is physically possible if the constant H also has the dimensions $ML^{-1}T^{-2}$.

1.2 Velocity of a pressure wave

The velocity of propagation a of a pressure wave through a liquid could be expected to depend upon the elasticity of the liquid represented by the bulk modulus K and its mass density ρ. Establish by dimensional analysis the form of a possible relationship.

Solution. Assume a simple exponential equation

$$a = CK^a \rho^b \tag{1}$$

where C is a numerical constant and a and b are unknown powers.

The dimensions of the variables are: velocity $a = LT^{-1}$, bulk modulus $K = ML^{-1}T^{-2}$, mass density $\rho = ML^{-3}$. If equation (1) is to be correct the powers a and b must be such that both sides of the equation contain the same powers of M, L and T. Rewrite equation (1) replacing each quantity by its dimensions, remembering that the constant C is a pure number.

$$LT^{-1} = M^a L^{-a} T^{-2a} \times M^b L^{-3b}$$

Equating powers of M, L and T,

$$0 = a + b$$
$$1 = -a - 3b$$
$$-1 = -2a$$

from which $\qquad a = +\tfrac{1}{2} \quad \text{and} \quad b = -\tfrac{1}{2}$

Thus a possible equation is $a = C\sqrt{\dfrac{K}{\rho}}$.

Compare this result with **10.5.**

Dimensional analysis gives the form of a possible equation but the value of the constant C would have to be determined experimentally.

1.3 Pipe flow

Show that a rational formula for the loss of pressure when a fluid flows through geometrically similar pipes is

$$p = \frac{\rho l v^2}{d} \phi\left(\frac{v d \rho}{\mu}\right)$$

where d is the diameter of the pipe, l is the length of the pipe, ρ is the mass density and μ the dynamic viscosity of the fluid, v is the mean velocity of flow through the pipe and ϕ means "a function of".

Solution. Assume $p = C\rho^a l^b v^c d^e \mu^f$ where C is a numerical constant and a, b, c, e, f are unknown powers.

The dimensions of the quantities are: $p = ML^{-1}T^{-2}$, $\rho = ML^{-3}$, $l = L$, $v = LT^{-1}$, $d = L$ and $\mu = ML^{-1}T^{-1}$.

Substituting these dimensions for the quantities,

$$ML^{-1}T^{-2} = M^a L^{-3a} \times L^b \times L^c T^{-c} \times L^e \times M^f L^{-f} T^{-f}$$

Equating powers of M, L and T,

$$1 = a + f \tag{1}$$

$$-1 = -3a + b + c + e - f \tag{2}$$

$$-2 = -c - f \tag{3}$$

There are five unknown powers and only three equations, so that it must be decided to solve for three of the unknown powers in terms of the others. In practice this decision is made from experience; in examination problems some indication is usually given in the question as to the form of the final result which depends on the choice of unknowns to be solved. In this case solve for the powers of ρ, v and d, namely a, c and e.

From equation (1) $a = 1 - f$

From equation (3) $c = 2 - f$

From equation (2) $e = -1 + 3a - c - b + f$

$\qquad\qquad\qquad = -f - b$

Substituting these values in the original equation

$$p = C\rho^{1-f} l^b v^{2-f} d^{-f-b} \mu^f$$

$$= C\rho v^2 \left(\frac{l}{d}\right)^b \left(\frac{\rho v d}{\mu}\right)^{-f}$$

$$= \frac{\rho v^2 l}{d} C \left(\frac{l}{d}\right)^{b-1} \left(\frac{\rho v d}{\mu}\right)^{-f}$$

For geometrically similar pipes $\dfrac{l}{d}$ is a constant and $\left(\dfrac{l}{d}\right)^{b-1}$ can be combined with C. Putting $K = C\left(\dfrac{l}{d}\right)^{b-1}$

$$p = \frac{\rho l v^2}{d} K \left(\frac{\rho v d}{\mu}\right)^{-f}$$

Since neither K nor f are known this is written simply as

$$p = \frac{\rho l v^2}{d} \phi \left(\frac{\rho v d}{\mu}\right) \tag{4}$$

It is interesting to compare this result with the Darcy formula

$$h_f = 4f\frac{l}{d}\frac{v^2}{2g}$$

From equation (4) $h_f = \dfrac{p}{\rho g} = \dfrac{l v^2}{dg} \phi \left(\dfrac{\rho v d}{\mu}\right)$

which indicates that the Darcy coefficient f must be a function of the pipe Reynolds number $\rho v d/\mu$. This has already been shown by more orthodox methods (*see* Volume 1).

1.4   Pipe flow

A rational formula for loss of pressure when fluid flows through geometrically similar pipes is

$$p = \frac{\rho l v^2}{d} \phi \left(\frac{\rho v d}{\mu}\right)$$

The measured loss of head in a 50 mm diam pipe conveying water at 0·6 m/s is 800 mm of water per 100 m length. Calculate the loss of head in millimetres of water per 400 m length when air flows through a 200 mm diam pipe at the "corresponding speed". Assume that the pipes have geometrically similar roughness and take the densities of air and water as 1·23 and 1000 kg/m³ and the absolute viscosities as 1·8 × 10⁻⁴ and 1·2 × 10⁻² poise respectively.

Solution. The formula $p = \dfrac{\rho l v^2}{d} \phi \left(\dfrac{\rho v d}{\mu}\right)$, derived by dimensional analysis in example **1.3**, might appear to be of little use since the nature of the function $\phi(\rho v d/\mu)$ is unknown, but it can be used for comparison of the pressure drops in two geometrically similar pipes provided that the value of the Reynolds number $\rho v d/\mu$ is the same in both cases. Then

$$\phi \left(\frac{\rho_1 v_1 d_1}{\mu_1}\right) = \phi \left(\frac{\rho_2 v_2 d_2}{\mu_2}\right)$$

and the ratio of pressure drops simplifies to

$$\frac{p_1}{p_2} = \frac{\rho_1}{\rho_2} \cdot \frac{l_1}{l_2} \cdot \frac{v_1^2}{v_2^2} \cdot \frac{d_2}{d_1}$$

The velocity of flow in the second pipe required to make the Reynolds number the same in both is known as the *corresponding speed*. Using the suffix w for the pipe containing water and a for that containing air, for equality of Reynolds numbers,

$$\frac{\rho_w v_w d_w}{\mu_w} = \frac{\rho_a v_a d_a}{\mu_a}$$

Corresponding speed for air $= v_a = v_w \, \dfrac{\rho_w}{\rho_a} \, \dfrac{d_w}{d_a} \, \dfrac{\mu_a}{\mu_w}$

$$= 0{\cdot}6 \, \frac{1000}{1{\cdot}23} \times \frac{50}{200} \times \frac{1{\cdot}8 \times 10^{-4}}{1{\cdot}2 \times 10^{-2}} = 1{\cdot}83 \, \text{m/s}$$

Ratio of pressure drops $\dfrac{p_a}{p_w} = \dfrac{\rho_a}{\rho_w} \dfrac{l_a}{l_w} \dfrac{d_w}{d_a} \dfrac{v_a^2}{v_w^2}$

$$= \frac{1{\cdot}23}{1000} \times \frac{400}{100} \times \frac{50}{200} \times \frac{1{\cdot}83^2}{0{\cdot}6^2} = 0{\cdot}01144$$

If loss of head per 100 m in 50 mm pipe is 800 mm of water
Loss of head per 400 m in 200 mm pipe $= 0{\cdot}01144 \times 800$

$$= \textbf{9·15 mm of water}$$

1.5 Resistance to a partially-submerged body

Find by dimensional analysis a rational formula for the resistance to motion R of geometrically similar bodies moving partially submerged through a viscous, compressible fluid of density ρ and coefficient of dynamic viscosity μ with a uniform velocity V.

Solution. The resistance R will be due to skin friction, wave resistance and compressibility of the fluid and will depend on the size of the body denoted by a characteristic length l, the velocity V, the density ρ, viscosity μ and bulk modulus K of the fluid and the gravitational acceleration g (for wave resistance). Thus R is a function of l, V, ρ, μ, K and g. The form of this function may be simple as was assumed in example **1.4** or may consist of a series of terms made up of the product of the variables each raised to suitable powers

$$R = Al^x V^y \rho^z \mu^p K^q g^r + A_1 l^{x_1} V^{y_1} \rho^{z_1} \mu^{p_1} K^{q_1} g^{r_1} + \ldots \tag{1}$$

where A, A_1, . . . are numerical constants, x, x_1, . . ., y, y_1, . . . etc. are unknown indices. Thus

$$\frac{R}{Al^x V^y \rho^z \mu^p K^q g^r} = 1 + \frac{A_1}{A} \, l^{x_1 - x} V^{y_1 - y} \rho^{z_1 - z} \mu^{p_1 - p} K^{q_1 - q} g^{r_1 - r}$$

Since the first term on the right-hand side is a pure number, the equation will only be correct if dimensionally

$$R = Al^x V^y \rho^z \mu^p K^q g^r$$

The dimensions of the quantities are: $R = MLT^{-2}$, $l = L$, $V = LT^{-1}$, $\rho = ML^{-3}$, $\mu = ML^{-1}T^{-1}$, $K = ML^{-1}T^{-2}$, $g = LT^{-2}$. Substituting in equation (1)

$$MLT^{-2} = L^z \times L^y T^{-y} \times M^z L^{-3z} \times M^p L^{-p} T^{-p} \times M^q L^{-q} T^{-2q} \times L^r T^{-2r}$$

Equating powers of M, L and T

$$1 = z + p + q \qquad (2)$$

$$1 = x + y - 3z - p - q + r \qquad (3)$$

$$-2 = -y - p - 2q - 2r \qquad (4)$$

Equations (2), (3) and (4) allow of three solutions only. A useful result is obtained by solving for x, y and z giving

$$z = 1 - p - q, \quad y = 2 - p - 2q - 2r, \quad x = 2 - p + r.$$

All the other terms on the right-hand side of equation (1) are similar to the first so that by the same dimensional reasoning

$$x_1 = 2 - p_1 + r_1, \quad y_1 = 2 - p_1 - 2q_1 - 2r_1, \quad z_1 = 1 - p_1 - q_1$$

and so on. Substituting in equation (1)

$$R = \rho V^2 l^2 \left\{ A \left(\frac{\rho V l}{\mu} \right)^{-p} \left(\frac{V}{\sqrt{(K/\rho)}} \right)^{-2q} \left(\frac{V}{\sqrt{(lg)}} \right)^{-2r} \right.$$
$$\left. + A_1 \left(\frac{\rho V l}{\mu} \right)^{-p_1} \left(\frac{V}{\sqrt{(K/\rho)}} \right)^{-2q_1} \left(\frac{V}{\sqrt{(lg)}} \right)^{-2r_1} + \dots \right\}$$

The series in brackets is an unknown function of $\dfrac{\rho V l}{\mu}$, $\dfrac{V}{\sqrt{(K/\rho)}}$ and $\dfrac{V}{\sqrt{(lg)}}$ and can be written

$$R = \rho V^2 l^2 \phi \left\{ \frac{\rho V l}{\mu}, \frac{V}{\sqrt{(K/\rho)}}, \frac{V}{\sqrt{(lg)}} \right\} \qquad (5)$$

The terms in the function are all dimensionless groups,

$\dfrac{\rho V l}{\mu}$ is the Reynolds number,

$\dfrac{V}{\sqrt{(K/\rho)}}$ is the Mach number and

$\dfrac{V}{\sqrt{(lg)}}$ is the Froude number.

Equation (5) may also be written

$$\frac{R}{\rho V^2 l^2} = \phi \left\{ \frac{\rho V l}{\mu}, \frac{V}{\sqrt{(K/\rho)}}, \frac{V}{\sqrt{(lg)}} \right\}$$

in which case $R/\rho V^2 l^2$ will also be found to be dimensionless.

1.6 Thrust of screw propeller

Assuming that the thrust F of a screw propeller is dependent upon the diameter d, speed of advance v, fluid density ρ, revolutions per second n and coefficient of viscosity μ, show that it can be expressed by the equation

$$F = \rho d^2 v^2 \phi \left\{ \frac{\mu}{\rho d v}, \frac{dn}{v} \right\}$$

Solution. F will be a function of d, v, ρ, n and μ. Instead of expanding this function fully as in example **1.5**, since all the terms are similar we can write

$$F = \Sigma A d^m v^p \rho^q n^r \mu^s \tag{1}$$

where A is a numerical constant and m, p, q, r and s are unknown powers.

The dimensions of the variables are $F = MLT^{-2}$, $d = L$, $v = LT^{-1}$, $\rho = ML^{-3}$, $n = T^{-1}$, $\eta = ML^{-1}T^{-1}$.

Substituting the dimensions for the variables, equation (1) will be true if

$$MLT^{-2} = L^m \times L^p T^{-p} \times M^q L^{-3q} \times T^{-r} \times M^s L^{-s} T^{-s}$$

Equating powers of M $1 = q + s$

$$L \quad\quad 1 = m + p - 3q - s$$
$$T \quad -2 = -p - r - s$$

The equation given in the problem indicates that it is desirable to solve for m, p and q in terms of r and s.

$$q = 1 - s, \quad p = 2 - r - s$$
$$m = 1 - p + 3q + s = 2 + r - s$$

Substituting in equation (1)

$$F = \Sigma A d^{2+r-s} v^{2-r-s} \rho^{1-s} n^r \mu^s$$

Regrouping by powers

$$F = \Sigma A \rho d^2 v^2 \left(\frac{\mu}{\rho d v}\right)^s \left(\frac{dn}{v}\right)^r$$

which can be written

$$F = \rho d^2 v^2 \phi \left\{ \frac{\mu}{\rho d v}, \frac{dn}{v} \right\}$$

where ϕ means "a function of".

1.7 Buckingham's Pi theorem

State Buckingham's Π theorem and apply it to the problem of example **1.6**.

Solution. Buckingham's Π theorem states that if there are n variables in a problem and these variables contain m primary dimensions (for example M, L and T) the equation relating the variables will contain $n - m$ dimensionless groups. Buckingham referred to these dimensionless groups as Π_1, Π_2, etc., and the final equation obtained is

$$\Pi_1 = \phi(\Pi_2, \Pi_3, \ldots \Pi_{n-m})$$

Thus in example **1.5** there are seven variables with three primary dimensions so that the final equation

$$\frac{R}{\rho V^2 l^2} = \phi \left\{ \frac{\rho V l}{\mu}, \frac{V}{\sqrt{(K/\rho)}}, \frac{V}{\sqrt{(lg)}} \right\}$$

is formed of four dimensionless groups.

In the problem of example **1.6** there are six variables, F, ρ, d, v, η and n and three primary dimensions. The equation relating the variables will therefore be formed of $6 - 3 = 3$ dimensionless groups and will be

$$\Pi_1 = \phi(\Pi_2, \Pi_3)$$

The dimensionless groups can be formed as follows:

(1) Choose a number of variables equal to the number of primary dimensions and including all these dimensions, in this case F, ρ and v.

(2) Form dimensionless groups by combining the variables selected in (1) with each of the others in turn.

Combining F, ρ and v with d to form a dimensionless group:

$$\Pi_1 = \frac{F}{\rho v^2 d^2}$$

Combining F, ρ and v with n, $\quad \Pi_2 = \dfrac{Fn^2}{\rho v^4}$

Combining F, ρ and v with μ, $\quad \Pi_3 = \dfrac{F\rho}{\mu^2}$

Since $\qquad\qquad\qquad\qquad \Pi_1 = \phi(\Pi_2, \Pi_3)$,

$$\frac{F}{\rho v^2 d^2} = \phi\left(\frac{Fn^2}{\rho v^4}, \frac{F\rho}{\mu^2}\right) \tag{1}$$

The groups can be rearranged to obtain the desired form by cross-multiplying. Rewrite equation (1) as

$$\frac{F}{\rho v^2 d^2} = \left(\frac{Fn^2}{\rho v^4}\right)^a \left(\frac{F\rho}{\mu^2}\right)^b \times \text{constant}$$

Multiplying both sides by $\left(\dfrac{F}{\rho v^2 d^2}\right)^{-a-b}$ gives

$$\left(\frac{F}{\rho v^2 d^2}\right)^{1-a-b} = \left(\frac{Fn^2}{\rho v^4} \cdot \frac{\rho v^2 d^2}{F}\right)^a \left(\frac{F\rho}{\mu^2} \cdot \frac{\rho v^2 d^2}{F}\right)^b \times \text{constant}$$

$$\left(\frac{F}{\rho v^2 d^2}\right)^{1-a-b} = \left(\frac{dn}{v}\right)^{2a} \left(\frac{\mu}{\rho v d}\right)^{-2b} \times \text{constant}$$

$$\frac{F}{\rho v^2 d^2} = \phi\left\{\frac{dn}{v}, \frac{\mu}{\rho v d}\right\}$$

Problems

1 Show that the frictional torque L required to rotate a disc of diameter d at an angular velocity ω in a fluid of viscosity μ and density ρ is given by

$$\frac{L}{d^5 \omega^2 \rho} = \phi\left\{\frac{\rho d^2 \omega}{\mu}\right\}$$

2 The rate of flow Q of a gas through a sharp-edged orifice depends on the diameter d of the orifice, the difference in pressure P between the two sides of the orifice, the density ρ and the kinematic viscosity v of the gas. Show by the method of dimensions that

$$Q = d^2 \left(\sqrt{\frac{P}{\rho}} \right) \phi \left\{ \frac{v}{d} \sqrt{\frac{\rho}{P}} \right\}$$

3 Show that the power P developed by a hydraulic turbine is given by

$$P = \rho N^3 D^5 \phi \left\{ \frac{N^2 D^2}{gH} \right\}$$

where ρ is the mass density of the fluid, N the speed of rotation, D the diameter of the rotor and H the available head.

4 Define viscosity and state the units in which it is measured. Show by applying the principle of dimensional homogeneity that the resistance to motion of a sphere through a viscous fluid is given by $R = K\mu dv$ where μ is the viscosity; d the diameter of the sphere; v the velocity; and K a numerical coefficient.

5 Assuming that for turbulent flow in a rough pipe the resistance per unit area of solid boundary τ is dependent upon the viscosity μ, the density ρ, the velocity V of the fluid, the diameter D of the pipe and the size of roughness k, show that

$$\frac{\tau}{\rho V^2} = \phi \left\{ \frac{VD\rho}{\mu}, \frac{k}{D} \right\}$$

6 In the rotation of similar discs in a fluid in which the motion of the fluid is turbulent, show by the method of dimensions that a rational formula for the frictional torque M of a disc of diameter D rotating at speed N in a fluid of viscosity μ and density ρ is

$$M = D^5 N^2 \rho \phi \left(\frac{\mu}{D^2 N\rho} \right)$$

7 Derive a general expression for the resistance to motion of a partially submerged body through a liquid in terms of the Froude and Reynolds numbers. How is this expression used to compare the resistance of a ship model with the full-size ship? Explain the assumptions usually made and quote experiments which justify them.

Describe in detail either (*a*) the production of a ship model; or (*b*) a method of measuring the resistance force when the model is towed through still water.

8 Describe, with the help of diagrams, the operation of a film lubricated journal bearing.

Specify the conditions necessary for strict geometrical similarity and show, by using the method of dimensions, that if temperature effects are neglected, the moment of frictional resistance for geometrically similar journal bearings can be expressed by

$$M = \mu N D^3 \phi \left(\frac{\mu N}{P} \right)$$

where μ is the viscosity of the lubricant, N is the speed of journal rotation, P is the load per unit projected area and D is the diameter of the journal.

Hence show that the moment of frictional resistance for all geometrically similar bearings running at "corresponding speeds" is proportional to PD^3.

9 Prove that the viscous resistance F of a sphere of diameter d, moving at constant speed v through a fluid of density ρ and viscosity μ, may be expressed as

$$F = \frac{\mu^2}{\rho} \phi \left(\frac{\rho v d}{\mu} \right)$$

Show that Stokes' result for low velocities, $F = 3\pi\mu v d$, is in agreement with this general formula.

A sample of emery powder was shaken in water contained in a glass beaker and then allowed to settle. It was found that the water cleared in 1 min 40 s when the depth was 18 cm. Calculate the minimum diameter of the particles, assuming them all spherical and taking the specific gravity of emery as $4 \cdot 0$ and the coefficient of viscosity of water as $0 \cdot 012$ poise.
Answer $0 \cdot 00364$ cm

10 Prove that the total resistance R to flow in a length l of a pipe of diameter d is given by

$$R = l d v^2 \rho \phi \left(\frac{\mu}{\rho v d} \right)$$

where ρ is the density of the fluid and v its mean velocity. From the above equation show that the loss of head h in a length l of a pipe can be expressed as $k l v^n / d^{3-n}$. Show that this can be applied to viscous flow if $n = 1$ and $k = 32\,\mu/pg$.

11 The discharge through a small orifice is dependent upon the head over the orifice H, the gravitational acceleration g, the diameter of the orifice D, the viscosity μ, the density ρ, the surface tension σ of the fluid and the roughness k. Find the dimensionless groups upon which the coefficient of discharge depends.

Answer $\dfrac{\rho D \sqrt{(gH)}}{\mu}, \dfrac{D}{H}, \dfrac{\sigma}{\rho g H^2}, \dfrac{k}{H}$

12 Show, by applying the method of dimensions, that a rational formula for the resistance of geometrically similar bodies moving, partially submerged, with uniform velocity V in a fluid having density ρ and viscosity μ is $R = \rho L^2 V^2 . \phi(N,F)$ where N denotes the Reynolds number and F denotes the Froude number. State the particular forms of resistance associated with each of these numbers; explain why dynamical similarity is not possible when applying this formula to predict full-scale resistance of ships from tests on models and describe how this difficulty is overcome.

13 Show that in the case of a cylindrical bearing without radial load, i.e. with uniform clearance such as a vertical bearing, the friction torque is given by

$$T = K \frac{\mu D^3 N L}{C} \text{ N-m}$$

where μ = dynamic viscosity of the oil (kg/m-sec); D = diameter of bearing (m); L = length of bearing (m); N = speed (rev/min); C = clearance (m); and find the value of the coefficient K.

Hence show by the method of dimensions that if the expression is to apply to similar journal bearings carrying radial load P, K must be replaced by $\phi(\mu D^2 N/P)$ where ϕ is an unknown function which can be obtained by experiment.

Describe how such an experiment would be carried out.
Answer $\pi^2/120$

14 Show, by the method of dimensions, that a rational formula for the flow over a vee-notch is given by

$$Q = g^{1/2} h^{5/2} \phi \left(\frac{g^{1/2} h^{3/2}}{\nu}, \frac{gh^2\rho}{\tau}, \theta \right)$$

where Q is the flow, h is the head above the vertex, ρ the density, v the kinematic viscosity, and τ the surface tension of the fluid, θ is the angle of the notch, and g the acceleration due to gravity.

Experiments with water on a 90° vee-notch show that a practical formula for flow is given by $Q = 1 \cdot 35 h^{2 \cdot 48}$, Q and h being in SI units.

Neglecting surface tension, estimate the percentage error involved if this formula is used for measuring the flow of oil whose kinematic viscosity is ten times that of water.
Answer 3·11 per cent

15 Show, by applying the method of dimensions, that the fall in pressure due to friction when an incompressible fluid having density ρ and viscosity μ flows through geometrically similar circular pipes can be expressed by

$$P = \frac{f' \rho L v^2}{d}$$

where d is the diameter and L is the length of the pipe, v is the mean velocity of flow and f' is a friction coefficient whose value depends upon the Reynolds number. State how all the quantities concerned are measured in SI units.

Show qualitatively by curves how the coefficient f' varies as the Reynolds number increases from about 1000 to 100 000 when the internal surface of the pipe is (*a*) smooth and (*b*) very rough. Give reasons for the shape taken by these curves in each case. Use a logarithmic base scale for the Reynolds number.

2

Dynamical similarity problems

So many factors are usually involved in any problem of fluid motion that it is often difficult to specify precisely the forces acting on a given fluid particle or the resultant motion. By making certain simplifying assumptions theoretical solutions can be obtained but these often require to be related to the practical results by a coefficient determined by experiment (as in the case of flow through an orifice). Dimensional analysis may also be used to find the possible form of a solution but again experiment is necessary to establish the precise relationship.

Where full-scale experiments cannot be undertaken, information can often be obtained from experiments on models provided that the relation between the results obtained from the model and the full-size counterpart can be established. This relationship will be simple only if the conditions in the model are such that the fluid flow is geometrically and dynamically similar to that in the counterpart.

2.1 General principles

> (a) What is meant by (1) geometrical similarity, (2) dynamical similarity?
>
> (b) What are the requirements for dynamical similarity between two fluid motions when considering (1) viscous resistance, (2) wave resistance, (3) compressibility, (4) surface tension?

Solution. (a) Two systems are geometrically similar when the ratio of corresponding lengths in the two systems is constant so that one is a scale model of the other.

The two systems are dynamically similar when the several forces acting on corresponding fluid elements have the same ratio to one another in both systems, so that the paths followed by the corresponding elements in the two systems will be geometrically similar. When a particle is set in motion its behaviour will depend on its inertia, which tends to carry it on at uniform velocity in a straight line, and upon the action of forces resisting motion. If the ratio of inertia to resisting force is the same in the two systems the two motions will be dynamically similar.

(b) The requirement in each case is that the ratio of inertia force to resisting force shall be the same in the two systems.

Let l be a characteristic length in the system under consideration, for example, the diameter of a pipe or the chord of an aerofoil, and t a typical time. Then the mass of an element is proportional to ρl^3 and its acceleration to l/t^2, so that

$$\text{Inertia force} \propto \text{mass} \times \text{acceleration}$$
$$\propto \rho l^3 \times l/t^2$$
$$\propto \rho l^2 \times (l/t)^2$$

or, since velocity $v \propto l/t$,

$$\text{Inertia force} \propto \rho l^2 v^2$$

(1) If the motion is controlled by viscous resistance, the flow in two systems will be dynamically similar if the ratio inertia force/viscous force is the same.

$$\text{Viscous force} \propto \text{viscous shear stress} \times \text{area}$$
$$\propto \mu \times \text{velocity gradient} \times l^2$$

and since velocity gradient $\propto v/l$

$$\text{Viscous force} \propto \mu v l$$

Thus $\qquad \dfrac{\text{inertia force}}{\text{viscous force}} \propto \dfrac{\rho v^2 l^2}{\mu v l} \propto \dfrac{\rho v l}{\mu}$

which is the Reynolds number.

Thus for viscous resistance the requirement for dynamical similarity in the two systems is equality of Reynolds numbers.

(2) For wave resistance the resisting force is due to gravity.

$$\text{Gravity force} = \text{mass} \times \text{gravitational acceleration}$$
$$\propto \rho l^3 \times g$$

Thus $\qquad \dfrac{\text{inertia force}}{\text{gravity force}} \propto \dfrac{\rho l^2 v^2}{\rho l^3 g} \propto \dfrac{v}{\sqrt{(lg)}}$

which is the Froude number. Two systems will be dynamically similar for wave resistance if the Froude number is the same for both.

(3) For elastic compression of a fluid the elastic force depends on the bulk modulus K of the fluid.

$$\text{Elastic force} \propto K l^2$$

Thus $\qquad \dfrac{\text{inertia force}}{\text{elastic force}} \propto \dfrac{\rho v^2 l^2}{K l^2} \propto \dfrac{v}{\sqrt{(K/\rho)}}$

which is the Mach number, so that for compressibility effects dynamical similarity is obtained when the Mach number is the same in both systems.

(4) For surface tension effects if T is the surface tension per unit length, surface tension force $\propto Tl$ and

$$\dfrac{\text{inertia force}}{\text{surface tension force}} \propto \dfrac{\rho v^2 l^2}{Tl}$$

which depends on $\dfrac{\rho v^2 l}{T}$ the Weber number.

2.2 Flow in a duct

> By dimensional analysis it can be shown that one form of the law governing the flow of fluids is
>
> $$\frac{F}{\rho v^2} = \phi\left(\frac{\rho v d}{\mu}\right)$$
>
> where F = frictional force per unit wetted area, ρ = mass density, v = velocity, d = pipe diameter, and μ = coefficient of viscosity.
>
> The flow of a gas in a uniform duct is to be simulated by means of water flow in a $\frac{1}{4}$-scale transparent model. The full-scale gas velocity is expected to be 24 m/s. Find: (a) the corresponding water velocity in the model; (b) the pressure drop to be expected per unit length of the full-scale duct if the pressure drop per unit length of the model is $13\cdot 8$ kN/m².
>
> 1 kg of gas under conditions in the full-scale duct occupies $0\cdot 686$ m³. μ for water is 62 times μ for gas.

Solution. Since the resistance is due to viscosity the requirement for dynamical similarity is that the Reynolds number $\rho v d/\mu$ must be the same in the model and the full-scale duct. Indicating values for gas by the suffix g and for water by the suffix w

$$\frac{\rho_g v_g d_g}{\mu_g} = \frac{\rho_w v_w d_w}{\mu_w}$$

(a) For equality of Reynolds numbers in the two systems,

$$\text{Corresponding water velocity } v_w = v_g \frac{\rho_g}{\rho_w}\cdot\frac{d_g}{d_w}\cdot\frac{\mu_w}{\mu_g}$$

Putting $v_g = 24\,\text{m/s}$, $\rho_g = \dfrac{1}{0\cdot 686}$ kg/m³, $\rho_w = 1000$ kg/m³

$$\mu_w/\mu_g = 62 \text{ and for a } \tfrac{1}{4}\text{-scale model } \frac{d_g}{d_w} = 4$$

$$v_w = 24 \times \frac{1}{0\cdot 686 \times 1000} \times 4 \times 62 = 8\cdot 68\,\text{m/s}$$

(b) Since the Reynolds number will be the same for both systems

$$\frac{F_g}{\rho_g v_g{}^2} = \phi\,(\text{Reynolds number}) = \frac{F_w}{\rho_w v_w{}^2}$$

$$F_g = F_w \frac{\rho_g}{\rho_w}\cdot\left(\frac{v_g}{v_w}\right)^2$$

If p = pressure drop in a pipe of length L and diameter d then since the velocity v is constant

Force due to pressure drop = frictional resistance

$$p \times \tfrac{1}{4}\pi d^2 = F \times \pi dL$$

$$p = \frac{4F}{d} L$$

Considering unit length of pipe in the two systems

$$\frac{p_g}{p_w} = \frac{F_g}{F_w} \times \frac{d_w}{d_g} = \frac{\rho_g}{\rho_w} \cdot \left(\frac{v_g}{v_w}\right)^2 \cdot \frac{d_w}{d_g}$$

Pressure drop/unit length in full-scale duct is

$$p_g = p_w \frac{\rho_g}{\rho_w} \left(\frac{v_g}{v_w}\right)^2 \frac{d_w}{d_g}$$

$$= 13 \cdot 8 \times \frac{1}{0 \cdot 686 \times 1000} \times \left(\frac{24}{8 \cdot 68}\right)^2 \times \frac{1}{4}\ \text{kN/m}^2$$

$$= 0 \cdot 0386\ \text{kN/m}^2$$

2.3 Drag on a sphere

A sphere of certain dimensions when placed in water moving with a velocity of $1 \cdot 5$ m/s experiences a drag of $4 \cdot 5$ N. Another sphere of twice the diameter is placed in a wind tunnel. Find the velocity of air for dynamical similarity. What will be the drag at this speed if the kinematic viscosity of air is 13 times that of water and the density of air is $1 \cdot 28$ kg/m³?

Solution. By dimensional analysis

$$\text{Drag on sphere} = F = \rho v^2 d^2 \phi\, \frac{vd}{\nu}$$

where ρ = mass density, v = fluid velocity, d = diameter, ν = kinematic viscosity of the fluid.

For dynamical similarity between the two spheres the Reynolds numbers must be the same for both systems. Using suffix a for air and w for water

$$\frac{v_a d_a}{\nu_a} = \frac{v_w d_w}{\nu_w}$$

Velocity of air for dynamical similarity $= v_a = v_w \dfrac{d_w}{d_a} \dfrac{\nu_a}{\nu_w}$

$$= 1 \cdot 5 \times \frac{1}{2} \times \frac{13}{1} = 9 \cdot 75\ \text{m/s}$$

At this velocity $\dfrac{v_a d_a}{\nu_a} = \dfrac{v_w d_w}{\nu_w}$

so that $\dfrac{F_a}{F_w} = \dfrac{\rho_a v_a^2 d_a^2}{\rho_w v_w^2 d_w^2}$

$$F_a = 4 \cdot 5 \times \frac{1 \cdot 28}{1000} \times \left(\frac{9 \cdot 75}{1 \cdot 5}\right)^2 \times \left(\frac{2}{1}\right)^2\ \text{N}$$

$$= 0 \cdot 976\,\text{N}$$

2.4 Drag in a compressible fluid

The resistance to motion F of a body moving through a compressible fluid of mass density ρ, dynamic viscosity and μ and bulk modulus K is given by

$$F = \rho v^2 l^2 \phi \left[\frac{\rho v l}{\mu}, \frac{v}{\sqrt{(K/\rho)}} \right]$$

where v is the velocity and l a characteristic length of the body.

From consideration of this relationship indicate the difficulty in obtaining dynamical similarity when testing scale models of high-speed aircraft in a wind tunnel.

How can the difficulty be overcome?

Solution. For dynamical similarity both the Reynolds number $\rho v l/\mu$ and the Mach number $v/\sqrt{(K/\rho)}$ must have the same values respectively in model and counterpart. Indicating quantities in the model by the suffix m, for equality of Reynolds numbers

$$\frac{\rho_m v_m l_m}{\mu_m} = \frac{\rho v l}{\mu}$$

If air is used for testing the model $\rho_m = \rho$ and $\mu_m = \mu$ so that the corresponding model speed is for equality of Reynolds numbers

$$v_m = v \times \frac{l}{l_m} = v\lambda \quad \text{where} \quad \lambda = \frac{l}{l_m}$$

For equality of Mach numbers

$$\frac{v_m}{\sqrt{(K_m/\rho_m)}} = \frac{v}{\sqrt{(K/\rho)}}$$

and using the same fluid $v_m = v$, since $K_m = K$ and $\rho_m = \rho$.

Equality of Reynolds numbers requires the model speed to be greater than full-size speed and equality of Mach numbers requires model and full-size speed to be the same. These requirements are incompatible when using air at atmospheric pressure as the tunnel fluid.

The difficulty can be overcome by using a compressed-air wind tunnel, thus making ρ_m greater than ρ while leaving μ substantially unchanged. Thus if the linear scale $\lambda = 10$ and the tunnel is pressurized to ten atmospheres

$$\rho_m = 10\rho$$

and for equality of Reynolds numbers

$$v_m = v \cdot \frac{\rho}{\rho_m} \cdot \frac{l}{l_m} = v \cdot \frac{\rho}{10\rho} \cdot 10 = v$$

Also, since $K = \gamma p/\rho$ and ρ varies directly with p the value of K is unchanged by pressurization and equality of Mach numbers can be maintained when $v_m = v$.

2.5 Wave resistance

What is meant by corresponding speeds in model experiments?

For a hydroplane the resistance is mainly due to wave formation. If the scale of a model hydroplane is 1/25 and if its resistance at a speed of 6 m/s is 1·8 N, what will be the resistance of the large hydroplane at the corresponding speed?

Solution. The corresponding speed is the speed at which the model should be operated to give dynamically similar conditions to those in the prototype.

If resistance is mainly due to wave formation, conditions will be dynamically similar if the Froude numbers are equal.

$$\left(\frac{v}{\sqrt{(lg)}}\right)_{\text{model}} = \left(\frac{V}{\sqrt{(Lg)}}\right)_{\text{full size}}$$

Corresponding speed of full-size ship is

$$v\sqrt{\frac{L}{l}} = 6\sqrt{\frac{25}{1}} = 30\,\text{m/s}$$

By dimensional analysis

$$\text{Wave resistance} = \rho L^2 V^2 \phi \text{ (Froude number)}$$

For dynamically similar conditions

$$\frac{\text{Resistance of full-size ship}}{\text{Resistance of model}} = \left(\frac{L}{l}\right)^2 \left(\frac{V}{v}\right)^2$$

$$\text{Resistance of full-size ship} = 1\cdot8 \times \left(\frac{25}{1}\right)^2 \left(\frac{30}{6}\right)^2$$

$$= 28\cdot15\,\text{N}$$

2.6 Ship resistance

A ship has a length of 132 m and a wetted area of 2325 m². A model of this ship, 4·2 m in length, is towed through fresh water at 1·5 m per s and found to have a total resistance of 17·75 N. The ship in sea water has an assumed skin resistance of 43 N/m² at 3 m per s and its velocity index is 1·85. For the model in fresh water the corresponding figures are 16 N/m² at 3 m per s and 1.9.

Calculate: (*a*) the corresponding speed of the ship, proving any formula used, and stating the assumptions made; (*b*) the shaft power required to propel the ship at this speed through sea water of density 1025 kg/m³, assuming a propeller efficiency of 70 per cent.

Solution. The total resistance of a ship is the sum of skin resistance and wave resistance. Sufficient information is given to calculate skin resistance for the model and for the ship. The purpose of the test is therefore to estimate the wave resistance of the ship and it must be carried out under conditions for dynamical similarity for wave resistance. By dimensional analysis

$$\text{Wave resistance } R = \rho l^2 v^2 \phi(F)$$

where ρ = fluid mass density, l = length, v = velocity and F = Froude number = $v/\sqrt{(lg)}$.

If v_m = model velocity, l_m = length of model, V = ship velocity, L = length of ship,

(a) For dynamical similarity

$$\frac{v_m}{\sqrt{(l_m g)}} = \frac{V}{\sqrt{(Lg)}}, \qquad \frac{V}{v_m} = \sqrt{\frac{L}{l_m}}$$

Corresponding speed of ship $V = v_m \sqrt{\frac{L}{l_m}}$

$$= 1 \cdot 5 \sqrt{\frac{132}{4 \cdot 2}} = 8 \cdot 4 \, \text{m/s}$$

(b) Total resistance = skin resistance + wave resistance
Skin friction/unit area of model = $q_m = k_m v_m^{1 \cdot 9}$, $q_m = 16 \, \text{N/m}^2$ when $v = 3 \, \text{m/s}$, so that

$$k_m = \frac{16}{3^{1 \cdot 9}} = \frac{16}{8 \cdot 06} = 1 \cdot 99$$

Linear scale of model = $l_m/L = 4 \cdot 2/132$

Wetted area of model $A_m = A \left(\frac{l_m}{L} \right)^2 = 2325 \times \left(\frac{4 \cdot 2}{132} \right)^2 = 2 \cdot 345 \, \text{m}^2$

Skin resistance of model at $1 \cdot 5 \, \text{m/s} = q_m A_m = k_m v_m^{1 \cdot 9} A_m$

$$= 1 \cdot 99 \times 1 \cdot 5^{1 \cdot 9} \times 2 \cdot 345 = 10 \cdot 1 \, \text{N}$$

Total resistance of model at $1 \cdot 5 \, \text{m/s} = 17 \cdot 75 \, \text{N}$
Wave resistance of model = $17 \cdot 75 - 10 \cdot 1 = 7 \cdot 65 \, \text{N}$
If R = wave resistance of ship and R_m = wave resistance of model under dynamically similar conditions

$$\frac{R}{R_m} = \frac{\rho}{\rho_m} \left(\frac{L}{l_m} \right)^2 \left(\frac{V}{v_m} \right)^2 = \frac{\rho}{\rho_m} \left(\frac{L}{l_m} \right)^3 \quad \text{since} \quad \frac{V}{v_m} = \sqrt{\frac{L}{l_m}}$$

Wave resistance of ship $= R_m \frac{\rho}{\rho_m} \left(\frac{L}{l_m} \right)^3$

$$= 7 \cdot 65 \times \frac{1025}{1000} \times \left(\frac{132}{4 \cdot 2} \right)^3 = 244500 \, \text{N}$$

Skin friction/unit area of ship $= q = kV^{1 \cdot 85}$

$$q = 43 \, \text{N/m}^2 \quad \text{when} \quad V = 3 \, \text{m/s}$$

so that $\qquad k = \dfrac{43}{3^{1 \cdot 85}} = \dfrac{43}{7 \cdot 46} = 5 \cdot 76$

Skin resistance of ship at $8.4\,\text{m/s} = qA = kV^{1.85}A$

$$= 5.76 \times 8.4^{1.85} \times 2325 = 686\,000\,\text{N}$$

Total resistance of ship = wave resistance + skin resistance
$$= 244\,500 + 686\,000 = 930\,500\,\text{N}$$

Work done/sec against resistance $= 930\,500 \times 8.4 = 7\,830\,000\,\text{W}$

$$\text{Shaft power} = \frac{100}{70} \times 7830 = \textbf{11\,200\,kW}$$

2.7   Vee-notch

The general expression for flow over a V-notch is given by

$$Q = H^2\sqrt{(gH)}\phi\left\{\frac{H\sqrt{(gH)}}{v}, \theta\right\}$$

in which H denotes the head, v is the kinematic viscosity of the liquid and θ the angle of the notch. Experiments on flow of water over a 90° V-notch gave the practical formula $Q = 1.37H^{2.47}\text{m}^3/\text{s}$ where H is the head in metres.

By applying the dynamical similarity conditions given by the general expression show that the percentage error involved in the use of the practical formula for measurement of the rate of flow in a liquid whose kinematic viscosity is 12 times that of water will be about 5 per cent and that the quantity will be under-estimated.

Solution. The general expression can be rewritten as

$$Q = A\left(\frac{H\sqrt{(gH)}}{v}\right)^n H^{2.5}$$

where A and n are unknown numbers and the practical formula is

$$Q = KH^{2.47}$$

For water the two formulae must agree so that

$$A\left(\frac{(\sqrt{g})H^{3/2}}{v_w}\right)^n H^{2.5} = KH^{2.47}$$

Comparing powers of H which must be the same on both sides

$$\frac{3}{2}n + 2.5 = 2.47$$

$$n = -0.02$$

For water $\qquad K_w = A(\sqrt{g})^{-0.02}v_w^{0.02}$

For the other liquid $\quad K_L = A(\sqrt{g})^{-0.02}v_L^{0.02}$

By the practical formula, for a given head H

$$\text{Estimated discharge } Q_{\text{est}} = K_wH^{2.47}$$

$$\text{True discharge } Q_{\text{act}} = K_L H^{2 \cdot 47}$$

$$Q_{\text{est}} = Q_{\text{act}} \times \frac{K_w}{K_L} = Q_{\text{act}} \frac{A(\sqrt{g})^{-0 \cdot 02} \nu_w^{0 \cdot 02}}{A(\sqrt{g})^{-0 \cdot 02} \nu_L^{0 \cdot 02}}$$

$$= Q_{\text{act}} \times \left(\frac{\nu_w}{\nu_L} \right)^{0 \cdot 02}$$

and putting $\nu_L = 12\nu_w$

$$\text{Estimated flow} = Q_{\text{act}} \times \left(\frac{1}{12} \right)^{0 \cdot 02} = \frac{1}{1 \cdot 05} \times \text{actual flow}$$

Flow is under-estimated by 5 per cent approx.

2.8   Tidal model

Explain why in making a model of a river or open channel it may be necessary to distort the model by using different vertical and horizontal scales of length, and show how the scales for other quantities are established.

A tidal model of a river estuary has a horizontal scale λ and a vertical scale μ. Derive expressions giving the ratio between: (a) the tidal period; (b) the rate of fall of silt, for the model and full-scale estuary.

Solution. River flow is normally turbulent but when the scale reduction for the model is large the value of Reynolds numbers in the model may be too low to ensure turbulent flow or, alternatively, there may not be sufficient operating head to allow the model to function. This can be overcome by distorting the model: (a) by tilting the model as a whole if the channel is nearly straight; (b) by using different vertical and horizontal scales to increase vertical difference of level; (c) by using models which are not geometrically similar but are hydraulic analogies.

If the horizontal scale (i.e., horizontal distance in model/horizontal distance full-size) k_x and the vertical scale k_z have been chosen, the scales for other quantities can be found from the equations relating the quantities.

Velocity $v = \sqrt{(2gh)}$ where $h = $ operating head and, since h is a vertical measurement, its scale ratio is k_z. The value of g will be the same in both systems ($k_g = 1$).

$$\text{Velocity scale } k_v = k_z^{1/2}$$

$$\text{Discharge } Q = \text{width} \times \text{depth} \times \text{velocity} = xzv$$

$$\text{Discharge scale } k_Q = k_x k_z k_v = k_x k_z^{3/2}$$

$$\text{Time } t = x/v. \text{ Time scale } k_t = k_x/k_v = k_x k_z^{-1/2}$$

The same process is used for other quantities.

(a) Time scale will determine the tidal period of the model.

$$\frac{\text{Time in estuary}}{\text{Time in model}} = \frac{\left(\dfrac{\text{distance}}{\text{velocity}}\right) \text{ in estuary}}{\left(\dfrac{\text{distance}}{\text{velocity}}\right) \text{ in model}} = \frac{x_1}{x_2} \times \frac{v_2}{v_1}$$

Since $v = \sqrt{(2gh)}$, $\quad \dfrac{v_2}{v_1} = \left(\dfrac{h_2}{h_1}\right)^{1/2}$

$$\frac{\text{Time in estuary}}{\text{Time in model}} = \frac{x_1}{x_2} \times \left(\frac{h_2}{h_1}\right)^{1/2} = \frac{\mu^{1/2}}{\lambda} \text{ where } \mu = \frac{x_2}{x_1} \text{ and } \lambda = \frac{h_2}{h_1}$$

(b) The rate of fall of silt must be such that a particle will sink from the free surface to the bed of the model in a time which represents to scale the time taken for a particle to fall in the full-size estuary.

$$\frac{\text{Rate of fall in estuary}}{\text{Rate of fall in model}} = \frac{\left(\dfrac{\text{depth}}{\text{time}}\right) \text{ in estuary}}{\left(\dfrac{\text{depth}}{\text{time}}\right) \text{ in model}}$$

$$= \frac{x_1}{x_2} \times \frac{t_2}{t_1} = \frac{1}{\mu} \times \frac{\lambda}{\mu^{1/2}} = \frac{\lambda}{\mu^{3/2}}$$

2.9 Surge tank

The diagram (Fig. 2.1) relates to a surge tank between a reservoir and a turbine. What is the purpose of such a tank?

Show that for complete and instantaneous closure of the turbine valve

$$dy = -\left(\frac{av}{A}\right) dt \quad \text{and} \quad y - Cv^2 = \frac{l}{g}\frac{dv}{dt}$$

where cv^2 is the loss of head by friction in a length l and A is the constant plan area of the tank.

Develop the law for comparing a model and full-scale tank:

$$\frac{R}{R_m} = \left(\frac{C_m}{C}\right)^2 \frac{l}{l_m} \left(\frac{v_m}{v}\right)^2$$

where the subscript m refers to the model and $R = A/a$.

Find a suitable diameter of model tank to satisfy the following conditions:

	l(m)	Pipe diameter (m)	Tank diameter (m)	C	Initial velocity (m/s)
Model	12	0·05	d	0·333	0·375
Prototype	180	1·20	3·00	0·333	1·500

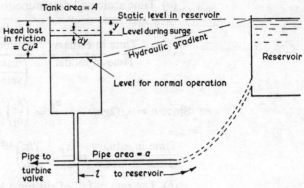

Figure 2.1

Solution. When a valve on a pipeline is suddenly closed, the pressure in the pipe will rise very rapidly to a value which far exceeds the static pressure and may cause damage to the pipe and valve (*see* Chapter 9). If a surge tank is provided close to the valve the flow in the pipe upstream of the surge tank is not suddenly brought to rest when the valve is shut and the pressure rise at the valve is reduced. As the flow passes into the surge tank the water level in the tank rises, reducing the head producing flow and bringing the flow in the main pipeline to rest gradually.

If v = velocity of flow in the pipeline at any instant t after closure of the valve when the level in the tank is y below reservoir level and if at time $t + \delta t$ the level in the tank changes to $y - \delta y$ below reservoir level then

$$-A\delta y = av\delta t$$

or in the limit

$$dy = -\left(\frac{av}{A}\right) dt \tag{1}$$

Also, if the velocity in the pipeline changes by δv in time δt

Rate of change of momentum = mass $\times$ rate of change of velocity

$$= \rho a l \frac{dv}{dt}$$

Force in direction of motion due to change of level in tank is

$$-\rho g \times (Cv^2 - y)a$$

By Newton's 2nd law, $-\rho g a (Cv^2 - y) = \rho a l \dfrac{dv}{dt}$

$$y - Cv^2 = \frac{l}{g}\frac{dv}{dt} \tag{2}$$

If $R = A/a$, equation (1) is

$$R = -\frac{v}{dy/dt}$$

Thus

$$\frac{R}{R_m} = \frac{v}{v_m} \frac{dt}{dt_m} \frac{dy_m}{dy}$$

or since the scale for dy = scale for y and the scale for dt = scale for t,

$$\frac{R}{R_m} = \frac{v}{v_m} \frac{t}{t_m} \frac{y_m}{y} \tag{3}$$

Also for equation (2) to be dimensionally correct the scales for each term must be the same so that

$$\frac{y}{y_m} = \frac{C}{C_m} \frac{v^2}{v_m^2}$$

and

$$\frac{l}{l_m} \frac{dV}{dV_m} \frac{dt_m}{dt} = \frac{lvt_m}{l_m v_m t} = \frac{C}{C_m} \frac{v^2}{v_m^2}$$

giving

$$\frac{t}{t_m} = \frac{l}{l_m} \frac{v_m}{v} \frac{C_m}{C}$$

Substituting in (3) for y/y_m and t/t_m

$$\frac{R}{R_m} = \frac{v}{v_m} \frac{lv_m C_m}{l_m vC} \frac{C_m v_m^2}{Cv^2}$$

$$\frac{R}{R_m} = \left(\frac{C_m}{C}\right)^2 \frac{l}{l_m} \left(\frac{v_m}{v}\right)^2$$

Substituting the values from the Table

$$R = \frac{\pi/4 \times 3 \cdot 00^2}{\pi/4 \times 1 \cdot 20^2} = 6 \cdot 25$$

$$R_m = 6 \cdot 25 \left(\frac{C}{C_m}\right)^2 \frac{l_m}{l} \left(\frac{v}{v_m}\right)^2$$

$$= 6 \cdot 25 \times 1 \times \frac{12}{180} \times \left(\frac{1 \cdot 5}{0 \cdot 375}\right)^2$$

$$= 6 \cdot 67 = \frac{A_m}{a_m} = \frac{d^2}{0 \cdot 05^2}$$

$$d^2 = 6 \cdot 67 \times 25 \times 10^{-4}$$

$$d = \mathbf{0 \cdot 129\,m}$$

Problems

1 In an experiment water flows through a 50 mm square pipe at 3·6 m/s and the loss of head is 940 mm in a length of 3 m. Find the corresponding speed for the flow of air in a duct 1 m square and calculate the loss of head in mm of water in a length of 90 m at that velocity. The density of air is 1·23 kg/m³ and the value of v for air is $1 \cdot 458 \times 10^{-5} m^2/s$ and for water $1 \cdot 180 \times 10^{-6} m^2/s$.
Answer 2·22 m/s, 0·662 mm of water

2 Water flowing in a certain 20 mm pipe becomes turbulent at a speed of 11·4 cm/s. What is the maximum speed for the flow of air to be laminar in a pipe 40 mm diam of similar construction? Take viscosity of water $= 1·12 \times 10^{-3}$ kg/m-s and of air as $17·7 \times 10^{-6}$ kg/m-s. Density of water $= 1000$ kg/m^3 and of air $= 1·23$ kg/m^3.
Answer 0·73 m/s

3 Define the terms "critical velocity" and "Reynolds number" as used in connexion with the flow of liquid through a pipe. If the critical velocity of water at 10°C in a 20 mm diam pipe is 0·13 m/s find the critical velocity of air at the same temperature in a similar pipe 0·3 m diam given that the kinematic viscosity of water at 10°C is $1·31 \times 10^{-6}$ m^2/s and of air $1·40 \times 10^{-5}$ m^2/s.
Answer 0·093 m/s

4 A bridge is supported by circular steel piles 0·3 m in diam. Find the force on each due to a stream 3·3 m deep running at 14·5 km/h given that the resistance per metre length of a long circular cylinder of 0·25 m diam tested in air is as follows

Speed (m/s)	23	36	46	57
Resistance (N/m)	23.7	78	152	248

For air $\rho = 1·23$ kg/m^3 $v = 1·48 \times 10^{-5}$ m^2/s
For water $\rho = 1000$ kg/m^3 $v = 1·31 \times 10^{-6}$ m^2/s
Answer 4·78 kN

5 An eighth-scale model of an aeroplane fuselage is tested in a high-pressure wind tunnel at 2150 kN/m^3 abs and 15°C. The coefficient of velocity is $18·1 \times 10^{-6}$ kg/m-s. Observed results are

Speed (m/s)	18	27	36
Resistance (N)	4.7	9.6	15.7

Find the full-scale resistance at 2400 m (at which pressure $=$ 75 kN/m^2 abs, temperature $= -1°C$ and $\mu = 17·1 \times 10^{-6}$ kg/m-s) at a speed of 290 km/h.
Answer 208 N

6 An orifice of diameter d is used to measure the volume rate of flow Q of a fluid having density ρ and coefficient of viscosity μ along a pipe of internal diameter D. The difference in pressure across the orifice is P. Show by dimensional analysis that the volume rate of flow is given by the expression

$$Q = Cd^2 \sqrt{\frac{P}{\rho}}$$

in which the coefficient C depends only on the value of the criterion $(\mu/\rho^{1/2} dP^{1/2})$ for all geometrically similar pipe arrangements in which (d/D) is constant.

A plate with a 90 mm diam orifice was fitted into a 250 mm diam pipeline to measure the rate of flow of water and the difference in pressure head across the orifice was 820 mm of mercury.

If the flow of air in a 100 mm diam pipe is to be dynamically similar calculate (1) the difference in pressure across the orifice in N/m^2, (2) the ratio of the volume rates of flow. Assume that for water $\rho = 1000$ kg/m³, $\mu = 15 \times 10^{-3}$ poise and for air $\rho = 1\cdot28$ kg/m³ and $\mu = 1\cdot8 \times 10^{-4}$ poise.

Answer 7693 N/m^2, $Q_a/Q_w = 3\cdot747$

7 Show by dimensional analysis that the power P required to operate a test tunnel is given by

$$P = \rho L^2 v^3 \phi \left\{ \frac{\mu}{\rho L v} \right\}$$

where ρ is the density and μ the coefficient of dynamic viscosity of the fluid, v the linear velocity of the fluid relative to the tunnel and L is a characteristic linear dimension of the tunnel.

A water tunnel was constructed for visual observation of the flow past models. It operates with the water flowing at a velocity of 3 m/s in the working section and absorbs $3\cdot75$ kW. If it is to operate as a wind tunnel under dynamically similar conditions, determine (*a*) the "corresponding speed" of air in the working section, and (*b*) the power required.

Assume that for water $\rho = 1000$ kg/m³ and kinematic viscosity $v = 1\cdot14 \times 10^{-6}$ m²/s, for air $\rho = 1\cdot28$ kg/m³, $v = 14\cdot8 \times 10^{-6}$ m²/s.

Answer $38\cdot9$ m/s, $10\cdot5$ kW

8 A ship having a hull length of 135 m is to be propelled at a speed of 30 km/h. Compute the Froude number. At what velocity should a 1 : 30 scale model be towed through water if the Froude number is to be the same for the model as for the prototype?

Answer $0\cdot229$, $1\cdot52$ m/s

9 Show that the wave-making resistance of a ship's hull may be expressed as

$$R = \rho v^2 L^2 \phi \left(\frac{v}{\sqrt{(gL)}} \right)$$

A ship is to be designed to cruise at $12\cdot5$ m/s. At the "corresponding speed" the wave-making component of resistance of a model to 1/40 scale is found to be 16 N when tested in fresh water. Find this model speed and also the wave-making resistance of the prototype in sea water weighing 1025 kg/m³.

Answer $1\cdot98$ m/s, 1050 kW

10 Show that the resistance to motion R of a surface vessel may be expressed as

$$R = \rho l^2 v^2 \phi \left(\frac{v}{vl}, \frac{lg}{v^2} \right)$$

where ρ is the density and v the kinematic viscosity of the liquid, l represents a dimension of the vessel, v is the velocity of the vessel, g the acceleration due to gravity and ϕ means "a function of".

A motor vessel, 9 m long and driven by a reaction jet, travels at 320 km per hour in such a manner that the viscous resistance is negligible compared with the wave-making resistance. A similar model of the vessel 150 mm long is found to require a force of 0·093 N to drive it over the water at the corresponding speed. What is the corresponding speed and what thrust is exerted by the jet? Assume that air resistance may be neglected.

Answer 41·3 km/h, 20 118 N

11 A one-sixth-scale model of a vessel is tested in a wind tunnel using air at 240 kN/m² and 21°C of density 28·5 kg/m³ and viscosity $18·39 \times 10^{-6}$ kg/m-s with a speed of 36·6 m/s. Calculate the corresponding speed of the full-scale vessel when submerged in sea water of density 1025 kg/m³ and having a viscosity of $1·637 \times 10^{-3}$ kg/m-s and find the drag if the model resistance is 67 N.

Answer 15·2 m/s, 15 000 N

12 Write a brief account of Froude's method for calculating the total resistance to motion of a ship from towing experiments on her model. Fundamental equations must be specified but need not be derived.

A vessel of 3170 metric tons has an indicated power of 1900 kW on trial at 28 km/h. Considering only resistance due to gravitational effects, calculate the probable power and corresponding speed of a vessel of the same form of 9150 metric tons displacement.

Answer 6536 kW

13 A model of a ship made to 1/20 scale, with wetted area 4 m², is towed in fresh water at 1·2 m/s. The measured resistance is 46·2 N. From tests on a thin plate of the same length as the model, the skin resistance is found to be $5·22 \, v^{1·95}$ n/m² of surface. Assuming skin resistance of the full-scale ship to be $4·95 \, v^{1·9}$ N/m² find the total resistance of the ship in sea water at the speed corresponding to the model speed of 1·2 m/s. Density of sea water 1025 kg/m².

Answer 327·5 kN

14 A ship is to have a length of 122 m and a wetted area of 2135 m². A model of it, 3·66 m in length, when towed at 1·3 m/s through fresh water had a total resistance of 15·3 N. Assuming the model has a skin resistance of 14·33 N/m² at 3 m/s and the velocity index is 1·9, while the ship at sea is assumed to have a skin resistance of 43 N/m² at 3 m/s and velocity index 1·85, calculate (*a*) the corresponding speed of the ship, proving any formula used, (*b*) the net power required to propel the ship at this speed through sea water of density 1025 kg/m³.

Answer 7·51 m/s, 6530 kW

15 Experiments are to be made on a $\frac{1}{10}$-scale model of a weir of irregular shape to determine the discharge of its prototype under a head of 1·2 m. What head would you use in the model experiments and if the model discharge was 42·5 dm³/s what would be the corresponding discharge of the large weir?

Answer 0·12 m, 13·44 m³/s

16 A tidal model of a river estuary has a horizontal linear scale of 1 in 7296. If the tidal period on the model is to be 62·5 s corresponding to 12 h 56 min in nature, what would be the most suitable vertical scale of ordinates on the model? Derive the formula used for this calculation.

Answer   1 in 95·7

17 Explain why it is necessary in the investigation of river flow by means of models to exaggerate the vertical scale of the model.

A model is to be made of a section of a river the roughness (Bazin N) of the river banks having been estimated as 0·83 SI units. If the average hydraulic mean depth of the section of the river is 2·6 m and that of the model 36 mm and the horizontal scale of the model is to be 1/200, determine the value which Bazin's roughness coefficient N should be for the model. Assume that the Chezy formula applies and that the coefficient is given by

$$C = \frac{86·8}{1 + \dfrac{N}{\sqrt{m}}} \text{ SI units}$$

If the vertical scale of the model is 1/50 and 11·8 kg of water are collected from the model in 1 min, find the river flow to which this corresponds. If the formulae are used they must be derived.

Answer   0·289 SI units, 13·9 m³/s

18 Experiments are to be conducted on tidal flow in a model of a river estuary in which the time interval between successive high tides is T. The linear scale ratios for the model are 1 in 500 horizontally and 1 in 25 vertically. What would be the time interval to be measured on the model?

Answer $T/100$

19 Moody's formula for the head lost by friction in the passages of a turbine is

$$\frac{FLV^{1·92}}{M^{1·25}}$$

where F is a frictional coefficient, L and M are the length and hydraulic mean radius respectively of the passages and V the velocity.

If F is the same for a scale model show that

$$\frac{1 - E}{1 - e} = \left(\frac{d}{D}\right)^{0·25} \times \left(\frac{h}{H}\right)^{0·04}$$

where E and e are the efficiencies and D and d the diameters of the prototype and model respectively, H and h being the corresponding heads.

Explain why the efficiency of the model is usually less than that of a full-size machine.

3

Vortex motion and radial flow

Both vortex motion and radial flow are examples of two-dimensional flow in which the streamlines are not parallel. In vortex motion the curvature of the streamlines introduces the action of centrifugal force which must be counterbalanced by a pressure gradient in the fluid. For radial flow the movement of the fluid is inward to or outward from a centre.

There are four types of vortices:

(1) *Forced vortex*. The fluid rotates as a solid body with constant angular velocity (*see also* Volume 1).

(2) *Free cylindrical vortex*. The fluid moves along streamlines which are horizontal concentric circles and there is no variation of total energy with radius.

(3) *Free spiral vortex*. This is a combination of the free cylindrical vortex with radial flow.

(4) *Compound vortex*. The fluid rotates as a forced vortex at the centre and as a free vortex outside.

3.1 Two-dimensional flow theory: Forced vortex

Deduce an expression for the variation of energy across the streamlines in a horizontal plane in a fluid.

A U-tube containing water is rotated about a vertical axis situated between the two limbs at a distance of 375 mm from one limb and 75 mm from the other. Calculate the difference of level of the water in the two limbs if the speed of rotation is 60 rev/min.

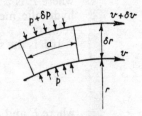

Figure   3.1

Solution. Consider an element of length *a* (Fig. 3.1), width δr and unit thickness lying between two horizontal streamlines of radius *r* and $r + \delta r$ which have velocities *v* and $v + \delta v$ respectively.

$$\text{Mass of element} = \rho a \delta r$$

$$\text{Centrifugal force} = \rho a \delta r \frac{v^2}{r}$$

This is counterbalanced by the force due to the pressure difference, so that

$$a \delta p = \rho a \delta r \frac{v^2}{r}$$

$$\frac{\delta p}{\rho} = v^2 \frac{\delta r}{r} \tag{1}$$

or in terms of pressure head

$$\text{Rate of change of pressure head radially} = \frac{dh}{dr} = \frac{v^2}{gr} \tag{2}$$

Also in moving from radius r to radius $r + \delta r$

$$\text{Rate of change of velocity head radially} = \frac{(v + \delta v)^2 - v^2}{2g\delta r}$$

$$= \frac{v}{g} \frac{\delta v}{\delta r}$$

neglecting products of small quantities.

Thus change of total energy H with radius is given by

$$\frac{dH}{dr} = \frac{v^2}{gr} + \frac{v}{g} \frac{dv}{dr} = \frac{v}{g} \left(\frac{v}{r} + \frac{dv}{dr} \right) \tag{3}$$

Note. It has been assumed that the streamlines are horizontal but equation (3) applies to cases where the streamlines are inclined to the horizontal also. The reason for this is that the element is in effect weightless since it is supported by the hydrostatic pressure of the surrounding fluid. If both the weight of the element and its inclination to the horizontal are considered in the calculation they will be found to cancel and the result will be identical with equation (3).

For a forced vortex the angular velocity ω is the same at all points.

$$v = \omega r \quad \text{so that} \quad \frac{dv}{dr} = \omega \quad \text{and} \quad \frac{v}{r} = \omega$$

$$\text{From equation (3)} \qquad \frac{dH}{dr} = \frac{\omega r}{g} \times 2\omega = \frac{2\omega^2 r}{g}$$

$$H = \frac{\omega^2 r^2}{g} + C \quad \text{where } C \text{ is a constant.}$$

But the total energy $H = \dfrac{p'}{\rho g} + \dfrac{v^2}{2g} + z$

$$= \frac{p}{\rho g} + \frac{\omega^2 r^2}{2g} + z$$

Thus
$$\frac{p}{\rho g} + \frac{\omega^2 r^2}{2g} + z = \frac{\omega^2 r^2}{g} + C$$

$$\frac{p}{\rho g} + z = \frac{\omega^2 r^2}{2g} + C$$

At the free surface of a forced vortex $p/\rho g = 0$ and the height z of the surface above datum is given by

$$z = \frac{\omega^2 r^2}{2g} + C \qquad (4)$$

Thus the free surface forms a paraboloid.

Figure 3.2 shows the rotating U-tube. The liquid in the tube forms

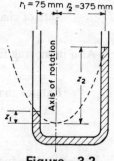

Figure 3.2

part of a forced vortex and will rise to levels z_1 and z_2 on the profile of the free surface of such a vortex (shown dotted). From equation (4)

$$z_2 - z_1 = \frac{\omega^2}{2g} (r_2{}^2 - r_1{}^2)$$

$$= \frac{(2\pi)^2}{2g} \{(0 \cdot 375)^2 - (0 \cdot 075)^2\} \, \text{m}$$

$$= 0 \cdot 272 \, \text{m}$$

3.2 Free vortex

Deduce an expression for the velocity at a given radius in a free vortex and hence obtain an equation for the pressure difference between two points on the same horizontal plane.

In such a vortex a point on the free surface at a radius of 150 mm is found to be 75 mm below the level of the free surface at the boundary of a vessel whose radius is very large. What will be the level of the surface at a radius of 300 mm below that at the boundary?

Solution. For a free cylindrical vortex the streamlines are concentric circles and there is no variation of the total energy with radius, i.e.

$$\frac{dH}{dr} = 0$$

From example **3.1**
$$\frac{dH}{dr} = \frac{v}{g}\left(\frac{v}{r} + \frac{dv}{dr}\right)$$

and so
$$\frac{v}{r} + \frac{dv}{dr} = 0$$

or
$$\frac{dv}{v} + \frac{dr}{r} = 0$$

Integrating,
$$\log_e v + \log_e r = \text{constant}$$

or $vr = C$ where C is a constant known as the *strength of the vortex*. At any radius r

$$v = \frac{C}{r} \tag{1}$$

Let p_1 and p_2 be the pressures in two concentric streamlines of radius r_1 and r_2 which have velocities v_1 and v_2. Since there is no change of total energy with radius, then for the same horizontal plane

$$\frac{p_1}{\rho g} + \frac{v_1^2}{2g} = \frac{p_2}{\rho g} + \frac{v_2^2}{2g}$$

$$\frac{p_1 - p_2}{\rho g} = \frac{v_2^2 - v_1^2}{2g}$$

From equation (1),
$$v_1 = \frac{C}{r_1} \quad \text{and} \quad v_2 = \frac{C}{r_2}$$

$$\frac{p_1 - p_2}{\rho g} = \frac{C^2}{2g}\left(\frac{1}{r_2^2} - \frac{1}{r_1^2}\right) \tag{2}$$

If the vortex is formed with a free surface then since there is no variation of total energy

$$z_1 + \frac{p_1}{\rho g} + \frac{v_1^2}{2g} = z_2 + \frac{p_2}{\rho g} + \frac{v_2^2}{2g} = H$$

At the free surface $p_1 = p_2$ and so

$$z_1 - z_2 = \frac{C^2}{2g}\left(\frac{1}{r_2^2} - \frac{1}{r_1^2}\right)$$

or alternatively
$$H - z = \frac{C^2}{2gr^2} \tag{3}$$

The free surface takes the form of a hyperbola asymptotic to the axis of rotation and to the horizontal through $z = H$ as shown in Fig. 3.3. In the present problem $H - z = 75\,\text{mm} = 0.075\,\text{m}$ when $r = 150\,\text{mm} = 0.15\,\text{m}$.

From equation (3) $\quad 0.075 = \dfrac{C^2}{2g \times (0.15)^2}\quad$ or $\quad\dfrac{C^2}{2g} = 0.001\,69$

When
$$r = 300\,\text{mm} = 0.3\,\text{m}$$

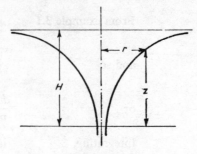

Figure 3.3

$$\text{Depression of surface} = H - z = \frac{C^2}{2g \times (0\cdot3)^2} = 0\cdot0188\,\text{m}$$

$$= \mathbf{18\cdot8\,mm}$$

3.3 Radial flow

Obtain an expression for the pressure difference between two points at radius r_1 and r_2 when a fluid flows radially inward or outward from a centre. Neglect frictional resistance.

Two stationary horizontal flat plates of external diameter $0\cdot375$ m are placed $12\cdot5$ mm apart. A vertical pipe 50 mm diam delivers $0\cdot41$ m³/min of water to the centre of the plates which are discharged at the periphery of the plates to atmosphere. The flow is radial and the pressure variation across the space between the plates at any radius can be ignored. Determine (*a*) the absolute pressure at the entrance to the plates and (*b*) the resulting thrust on the upper plates. Atmospheric pressure is 101 kN/m².

Solution. Flow is radial and therefore in straight lines so that r, the radius of curvature of the streamlines, is infinite, $dH/dr = 0$, and for all streamlines

$$H = \frac{p}{\rho g} + \frac{v^2}{2g} = \text{constant}$$

The arrangement is shown in Fig. 3.4. If p_1 and p_2 are the pressures at radius R_1 and R_2 respectively where the velocities are v_1 and v_2

$$\frac{p_1 - p_2}{\rho g} = \frac{v_2{}^2 - v_1{}^2}{2g}$$

For continuity of flow, $2\pi R_1 t v_1 = 2\pi R_2 t v_2 = Q$ where Q is the discharge,

or $\qquad\qquad v_1 = \dfrac{Q}{2\pi R_1 t} \quad \text{and} \quad v_2 = \dfrac{Q}{2\pi R_2 t}$

Thus $\qquad\qquad \dfrac{p_1 - p_2}{\rho g} = \dfrac{Q^2}{8\pi^2 t^2 g}\left(\dfrac{1}{R_2{}^2} - \dfrac{1}{R_1{}^2}\right) \qquad\qquad (1)$

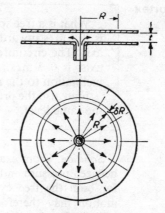

Figure 3.4

Compare this equation with equation (2) of example **3.2**. The pressure distribution has the same hyperbolic variation with radius.

In equation (1) put $Q = 0.41\,\text{m}^3/\text{min} = 0.00684\,\text{m}^3/\text{s}$, $t = 12.5\,\text{mm} = 0.0125\,\text{m}$, $R_1 = 0.025\,\text{m}$, $R_2 = 0.1875\,\text{m}$, $p_2 = 101 \times 1000\,\text{N/m}^2$, $p_1 = $ pressure at inlet.

$$\frac{p_1 - p_2}{\rho g} = \frac{(6.84 \times 10^{-3})^2}{8\pi^2 \times (12.5 \times 10^{-3})^2 g}\left\{\frac{1}{(0.1875)^2} - \frac{1}{(0.025)^2}\right\} \begin{matrix}\text{metres}\\\text{of water}\end{matrix}$$

$$= -0.607\,\text{m of water}$$

Absolute pressure at inlet $= p_1 = 101\,000 - 0.607 \times 9.81 \times 10^3 = \mathbf{95040\,N/m^2}$.

On any annulus of radius R and width δR resultant pressure intensity $= p_2 - p$ since pressure outside is the atmospheric pressure p_2.

$$p_2 - p = \frac{\rho g Q^2}{8\pi^2 t^2 g}\left(\frac{1}{R^2} - \frac{1}{R_2{}^2}\right)$$

Resultant force on the ring $= \dfrac{\rho g Q^2}{8\pi^2 t^2 g}\left(\dfrac{1}{R^2} - \dfrac{1}{R_2{}^2}\right) 2\pi R \delta R$

Integrating for the portion between $R = R_1$ and $R = R_2$, adding the force on the portion above the inlet pipe assuming a constant pressure p_1, and allowing for the force due to change of direction of flow,

Resultant force pushing plates together

$$= \frac{\rho Q^2}{4\pi t^2}\int_{R_1}^{R_2}\left(\frac{\delta R}{R} - \frac{R\delta R}{R_2{}^2}\right) + \pi R_1{}^2(p_2 - p_1) - \rho Q v_1$$

$$= \frac{\rho Q^2}{4\pi t^2}\left\{\log_e\frac{R_2}{R_1} - \frac{(R_2{}^2 - R_1{}^2)}{2R_2{}^2}\right\} + \pi R_1{}^2(p_2 - p_1) - \frac{\rho Q^2}{2\pi R_1 t}$$

$$= \frac{10^3 \times (6.84 \times 10^{-3})^2}{4\pi \times (12.5 \times 10^{-3})^2}\left\{\log_e 7.5 - \frac{(0.1875)^2 - (0.025)^2}{2 \times (0.1875)^2}\right\}$$

$$+ \pi \times (0.025)^2 \times 5960 - \frac{10^3(6.84 \times 10^{-3})^2}{2\pi \times 0.025 \times 12.5 \times 10^{-3}}\,\text{N}$$

$$= 23.9\,(2.015 - 0.472) + 11.72 - 23.8 = \mathbf{24.8N}$$

3.4 Free spiral vortex

What is a free spiral vortex? In such a vortex the radial velocity is 0·9 m/s inwards at a point A at a radius of 0·3 m from the axis. The circumferential velocity at a point B, 0·9 m from the axis, is 0·3 m/s. Calculate the resultant velocity at A and its inclination to the radius.

Find also the pressure difference between A and B if they are at the same level.

Solution. From examples **3.2** and **3.3** it will be seen that the relation between pressure and radius and between velocity and radius is similar for both the free cylindrical vortex and radial flow. Both types of motion may therefore occur together. The fluid rotates and flows radially forming a free spiral vortex in which a particle will follow a spiral path (Fig. 3.5).

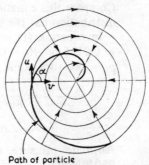

Path of particle

Figure 3.5

At any radius r let v = radial velocity

u = circumferential velocity

V = resultant velocity

From Example **3.3**, $\qquad vr$ = constant

From Example **3.2**, $\qquad ur$ = constant

Thus $\qquad v_A r_A = v_B r_B$

$$v_B = v_A \frac{r_A}{r_B} = 0\cdot9 \times \frac{0\cdot3}{0\cdot9} = 0\cdot3 \, \text{m/s}$$

and $\qquad u_A r_A = u_B r_B$

$$u_A = u_B \frac{r_B}{r_A} = 0\cdot3 \times \frac{0\cdot9}{0\cdot3} = 0\cdot9 \, \text{m/s}$$

Resultant velocity at A $= V = \sqrt{(v_A{}^2 + u_A{}^2)} = \sqrt{(0\cdot9^2 + 0\cdot9^2)}$

$$= 1\cdot273 \, \text{m/s}$$

If α = angle between streamline and radius

$$\tan \alpha = \frac{u_A}{v_A} = \frac{0\cdot9}{0\cdot9} = 1$$

$$\alpha = 45°$$

Since there is no energy change from A to B, by Bernoulli's theorem

$$\frac{p_A}{\rho g} + \frac{V_A^2}{2g} + z_A = \frac{p_B}{\rho g} + \frac{V_B^2}{2g} + z_B$$

Putting $z_A = z_B$, $V_A = \sqrt{(v_A^2 + u_A^2)}$ and $V_B = \sqrt{(v_B^2 + u_B^2)}$

$$\frac{p_B - p_A}{\rho g} = \frac{(u_A^2 + v_A^2) - (u_B^2 + v_B^2)}{2g}$$

$$= \frac{(0\cdot81 + 0\cdot81) - (0\cdot09 + 0\cdot09)}{2g} = 0\cdot0734\,\text{m head}$$

3.5 Compound vortex

Flow takes place along circular streamlines in a horizontal plane. If the pressure head is h and the tangential velocity is v along a streamline having radius r, show that the rate of change of h in a radial direction is given by

$$\frac{dh}{dr} = \frac{v^2}{gr}$$

A compound vortex produced in water having a free surface comprises a central forced vortex surrounded by a free vortex. The change in the type of vortex motion occurs at a radius of $0\cdot15$ m and the depth of the centre of the depression is $0\cdot6$ m below the free surface level, where $r \to \infty$.

Obtain the angular velocity of the forced vortex.

Solution. In example **3.1**, equation (2), it has been shown that

$$\frac{dh}{dr} = \frac{v^2}{gr}$$

The compound vortex is shown in Fig. 3.6. From example **3.1** for a forced vortex $v = \omega r$ and at any radius r the height of the free surface above the centre of the depression is given by $z = \omega^2 r^2/2g$. From example **3.2** for a free vortex $v = C/r$ and the depression of the surface at radius r below the surface at infinity is

$$d = \frac{C^2}{2gr^2} = \frac{v^2}{2g}$$

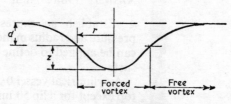

Figure 3.6

Taking r as the common radius of the two vortices

$$\text{Total depression} = z + d = \frac{\omega^2 r^2}{2g} + \frac{v^2}{2g}$$

Since at the common radius the velocities of the two vortices are equal, $v = \omega r$ so that

$$z + d = \frac{\omega^2 r^2}{g}$$

Putting $z + d = 0\cdot6\,\text{m}$, $r = 0\cdot15\,\text{m}$,

$$0\cdot6 = \frac{\omega^2 \times 0\cdot15^2}{g}$$

$$\omega = \frac{\sqrt{(0\cdot6g)}}{0\cdot15} = \mathbf{16\cdot17\,rad/s}$$

Problems

1 The outer and inner diameters of the impeller of a centrifugal pump are $1\cdot2$ m and $0\cdot6$ m respectively. Find the speed at which lifting commences against a head of 9 m.
Answer 244 rev/min

2 A tube ADBC consists of a straight vertical portion ADB and a curved portion BC, BC being a quadrant of a circle 255 m radius with centre at D. End A is open and the end C is closed. The tube is completely filled with water up to a height of 250 mm above B. Find the speed of rotation about the vertical axis ADB for the pressure C to be the same as the pressure at B. For this speed find the maximum pressure in BC and its position. Prove any formula used.
Answer $89\cdot2$ rev/min, $0\cdot195$ m rad., 3 kN/m²

3 In a hydraulic footstep bearing a flat circular disc 680 mm diam is attached to the lower end of a 100 mm diam vertical shaft. The upper face of the disc carries a series of radial ribs so that the water above it rotates with the shaft and disc. The space below the disc forms a pressure cylinder in which the water is kept at rest by ribs cast in the cylinder walls. The water in the upper and lower spaces is in free communication at the edge of the disc. Calculate the speed in rev/min at which the shaft and disc must rotate to balance an axial downward thrust of 4500 N.
Answer 198 rev/min

4 Establish an expression giving the relation between the pressure and radius in the case of a forced vortex and show how it can be applied to define the shape of a forced vortex with a free surface.

A cylindrical vessel $0\cdot6$ m diam and $0\cdot3$ m high is open at the top except for a lip 50 mm wide all round and normal to the side. It contains water to a height of $0\cdot2$ m above the bottom and is

rotated with its axis vertical. Calculate the speed in rev/min at which the water reaches the inside edge of the lip and the total pressure of water against the underside of the lip.
Answer 87·5 rev/min, 49·5 N/m²

5 An enclosed horizontal channel of square section, side *s* metres, has vertical sides and at one place there is a right-angled bend, the radius of the bend at the centreline of the channel being *R* metres. If the channel is filled by a liquid flowing along it, show that with a perfect fluid the product *vr* is constant, where *v* is the velocity of the fluid at radius *r*. Show also that for $s = \frac{1}{2}R$ the quantity flowing is $Q = 4 \cdot 244s^2\sqrt{h}$ m³/s, where *h* metres is the increase in pressure head in the bend between the inner and outer sides of the channel.

6 An impeller, 0·3 m diam, rotating concentrically about a vertical axis inside a closed cylindrical casing, 0·9 m diam, produces vortex motion in the water with which the casing is completely filled. Inside the impeller, the vortex motion is forced whilst, between the impeller and the circular side of the casing, the vortex is free. The pressure head at the side of the casing is 12 m of water when the pressure head at the centre of the casing is 1·2 m of water. Obtain the speed of the impeller in rev/min.
Answer 213 rev/min

7 Inward radial flow occurs between two horizontal discs 0·6 m in diam and 75 mm apart, the water leaving through a central pipe 150 mm diam in the lower disc at the rate of 0·17 m³/s. If the absolute pressure at the outer edge of the disc is 101 kN/m² calculate the pressure at the outlet. Find also the resultant force on the upper disc.
Answer 90 kN/m², 567 N

8 Two horizontal discs are 12·5 mm apart and 300 mm diam. Water flows radially outwards between the discs from a 50 mm diam pipe at the centre of the lower disc. If the pressure at the outer edge of the disc is atmospheric, calculate the pressure in the supply pipe when the velocity of the water in the pipe is 6 m/s. Find also the resultant pressure on the upper disc, neglecting impact force.
Answer −17·5 kN/m³, 126·7 N

9 A forced vortex 0·3 m diam is rotated at 120 rev/min and is surrounded by a free vortex of infinite extent. Find the height of the surface at infinity above the level at the centre.
Answer 0·367 m

10 A compound vortex formed in water having a free surface comprises a central forced vortex surrounded by a free vortex. The relation between the velocity in m/s and the radius in metres is $v = 10r$ for the forced vortex and $vr = 0 \cdot 9$ for the free vortex. Obtain the depth of the centre of the depression in the water surface below the free surface level.
Answer 0·92 m

4

Streamlines and stream function

The paths of fluid particles can be shown by drawing *streamlines* which are curves or straight lines such that the velocity of the fluid at any point is always tangential to the streamline. There is no flow across a streamline and the portion of fluid lying between any two streamlines can be considered in isolation. Thus in Fig. 4.1 for a unit thickness of

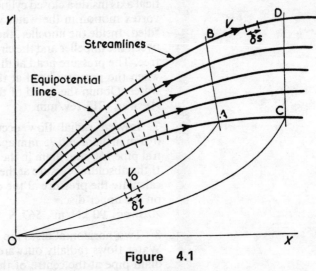

Figure 4.1

flow the cross-sectional area at CD is greater than at AB and since the same discharge passes through both sections the velocity is greater at AB than at CD. Diverging streamlines indicate falling velocity.

The position of streamlines can be calculated for the irrotational flow of an ideal inviscid fluid flowing in two dimensions and in certain cases the results are similar to flow patterns obtained experimentally.

Stream function

Let O, Fig. 4.1, be a reference point and A and C two points on a streamline.

Stream function of A = ψ = flow or flux across OA

$$= \int_0^A V_0 \delta l$$

where V_0 is the normal velocity across an element of OA of length δl.

For steady flow, the flow across OA = flow across OC since fluid is not appearing or disappearing in the area OAC and there is no flow across the streamline AC. The stream function ψ is therefore constant for any streamline.

The units of ψ will be those of velocity × length (m²/s) or discharge per unit thickness of flow, and ψ is a scalar quantity.

The numerical value of ψ for a streamline depends on the reference point and is assumed positive when flow is clockwise about O. Values of ψ increase positively to the left of the reference line (x-axis) when looking downstream.

Velocity potential

If V is the velocity at any point on a streamline and δs is a short distance along the streamline from that point then

$$\text{Velocity potential of D relative to B} = \phi = \int_B^D V ds$$

Points can be marked off along each streamline which will have the same value of ϕ and can be joined by equipotential lines which are always at right angles to the streamlines.

When the streamlines converge the velocity V increases and therefore for a given increment of ϕ the distance between the equipotential lines will also decrease.

In a real fluid subjected to shearing force the equipotential lines will cease to be perpendicular to the streamlines.

4.1 Stream function

> Explain the meaning of the term "stream function". An incompressible fluid is in steady motion past a curved surface. Measurements are made of the velocity of flow at a number of points along each of a series of normals to the surface. Show how these results may be used to plot streamlines.

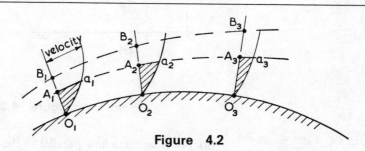

Figure 4.2

Solution. The term "stream function" is defined above.

In Fig. 4.2 O_1, O_2, O_3, etc., are the normals on which curves have been plotted showing the velocity measurements. Thus A_1a_1 represents the velocity at A_1 and the shaded area $O_1A_1a_1$ represents the flow across O_1A_1 which will be the stream function. If similar points A_2, A_3, etc., are found so that the areas $O_1A_1a_1 = O_2A_2a_2 = O_3A_3a_3 = \ldots$ then A_1, A_2, $A_3 \ldots$ lie on the same streamline since they will have the same stream function.

The process can be repeated for any other series of points B_1, B_2, B_3 and so on.

4.2 Relation to velocity

> For the flow of an incompressible inviscid fluid obtain a relationship between the stream function ψ and the components of the velocity u and v parallel respectively to the x- and y-axes.
>
> Obtain expressions for the stream functions for (a) a uniform flow of 20 m/s parallel to the positive direction of the x-axis, (b) a uniform flow of 10 m/s parallel to the negative direction of the y-axis, (c) a combination of (a) and (b).

Solution. In Fig. 4.3(a) let the stream function of A be ψ and of B be $\psi + \delta\psi$.

Flow across element (clockwise about A) $= \delta\psi$

$$= u\delta y - v\delta x$$

or in the limit $\qquad\qquad dy = udy - vdx$

But $\qquad\qquad\qquad dy = dy\,\dfrac{\partial\psi}{\partial y} + dx\,\dfrac{\partial\psi}{\partial x}$

Thus $\qquad\qquad u = +\dfrac{\partial\psi}{\partial y} \quad\text{and}\quad v = -\dfrac{\partial\psi}{\partial x}$

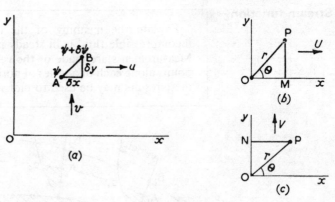

Figure 4.3

(a) For a uniform flow parallel to the x-axis with velocity U in the positive direction (Fig. 4.3(b)) considering any point P:

$$\psi = \text{flux across OP or since } y = 0 \text{ is a streamline}$$
$$\psi = \text{flux across PM} = +Uy = +Ur \sin\theta$$

In the present case $U = 20$ m/s and the horizontal lines in Fig. 4.4 represent to scale the stream function $\psi_1 = +20y$.

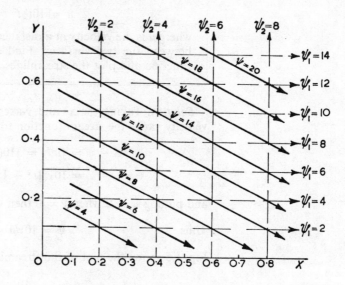

Figure 4.4 Dimensions X and Y are in metres

(b) For a uniform flow parallel to the y-axis with velocity V in the positive direction (Fig. 4.3(c)) considering any point P:

ψ = flux across OP = flux across PN since ON is a streamline

$$\psi = -Vx = -Vr\cos\theta$$

In the present case $V = -10$ m/s and the vertical lines in Fig. 4.4 represent to scale the stream function $\psi_2 = +10x$.

(c) Since stream functions are scalar quantities

Stream function of combined flow $\psi = \psi_1 + \psi_2$

$$\psi = 20y + 10x$$

The streamlines of the combined flow can be drawn on Fig. 4.4 by joining the intersections of ψ_1 and ψ_2 for which $\psi_1 + \psi_2 = $ constant.

Resultant velocity of the combined flow is given by a triangle of velocity as

$$q = \sqrt{(U^2 + V^2)}$$
$$= \sqrt{(20^2 + 10^2)} = \mathbf{22\cdot3\,m/s}$$

and is inclined at $\tan^{-1}(10/20) = \mathbf{26\tfrac{1}{2}°}$ to the x-axis.

4.3 Stream function for a parabolic velocity distribution

The velocity distribution between two flat surfaces a distance $t = 2$ m apart is given by

$$u = 10(\tfrac{1}{4}t^2 - y^2)\text{m/s}$$

where u is the velocity at a distance y from a plane lying midway between the two surfaces. Find an expression for the stream function and plot the streamlines.

Solution. Taking the x- and y-axes as shown in Fig. 4.5, since the velocity varies the stream function for any point from the axis will be

$$\psi = \int u\,dy = \int 10(\tfrac{1}{4}t^2 - y^2)dy$$
$$= 10y(\tfrac{1}{4}t^2 - \tfrac{1}{3}y^2) + C$$

and putting $\psi = 0$ when $y = 0$ then $C = 0$.

Thus
$$\psi = 10y(1 - \tfrac{1}{3}y^2)$$

Fig. 4.5 shows the plot of the streamlines.

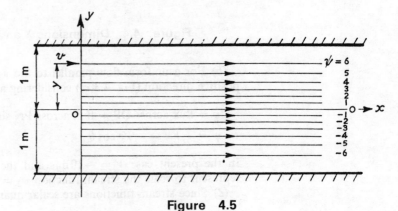

Figure 4.5

4.4 Sources and sinks

What are the meanings of the terms "point source" and "point sink" and how may their strengths be measured in two-dimensional flow?

Solution. A *source* is a point from which fluid flows out uniformly in all directions. A two-dimensional source is a line of unit length perpendicular to the plane of flow from which fluid flows along radiating streamlines.

A *sink* is a negative source, a point to which fluid flows uniformly from all directions and disappears.

Figure 4.6 shows a source O from which streamlines radiate in all directions.

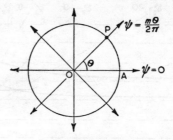

Figure 4.6

Strength of the source m = total output of the source

$$= \text{total flux across any circle with the source as centre.}$$

If u = velocity at any radius r

$$\text{Strength of source } m = 2\pi r u$$

The units are those of discharge per unit thickness (m²/s). The streamlines are radial and if OA is chosen as the streamline $\psi = 0$ then for the point P

$$\psi_P = \text{flux across AP} = \frac{m\theta}{2\pi}$$

By convention the angle θ in radians is limited to $\pm\pi$.

4.5 Uniform flow and sink

Establish the form of the stream function (*a*) for a uniform indefinite flow parallel to the x-axis in the direction from x positive towards the origin and (*b*) for a sink of strength m.

(*c*) Draw the streamlines resulting from a uniform flow as above over a sink, the scale for the uniform flow being 1 cm = 4 units of flow while the sink has a strength of 24 units. Draw the streamlines from $\psi = -20$ to $\psi = +20$ by movements of $\psi = 4$.

Solution. (*a*) From example **4.2** for a uniform flow from x positive towards the origin

$$\psi = -Uy$$

(*b*) From example **4.4** since a sink is a negative source

$$\psi = -\frac{m\theta}{2\pi}$$

(*c*) For the uniform flow $\psi_1 = -10y$

For the sink $\qquad \psi_2 = -24\dfrac{\theta}{2\pi}$

The streamlines corresponding to the combined flow are shown in Fig. 4.7 as the addition of the two flows, $\psi = \psi_1 + \psi_2$.

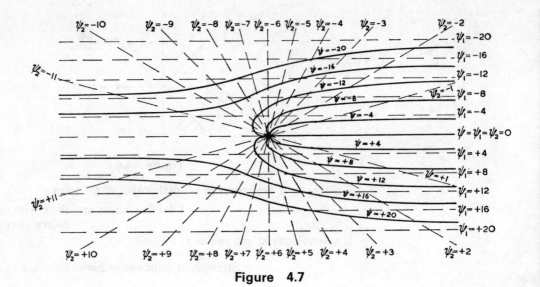

Figure 4.7

4.6 Uniform flow and source

A line source of strength m is placed at the origin in an otherwise uniform stream of an inviscid incompressible fluid of velocity $-U$ parallel to the x-axis. Write down the resulting stream function for the combined flow and determine the equation of the streamline which branches at the stagnation point. In particular determine in terms of m and U the maximum distance measured parallel to the y-axis between the branches. What is the value of the pressure coefficient on this streamline at the point where the y-axis cuts it?

Solution. For the uniform flow $\psi_1 = -Uy$

For the source $\psi_2 = +\dfrac{m\theta}{2\pi}$

For the combined flow $\psi = \psi_1 + \psi_2 = -Uy + \dfrac{m\theta}{2\pi}$

These streamlines are plotted in Fig. 4.8.

The streamline which branches at the stagnation point S is $\psi = 0$

for which $-Uy + \dfrac{m\theta}{2\pi} = 0$

$$y = m\theta/2\pi U$$

Note. Figure 4.8 represents the flow around a solid body whose boundary is the streamline $\psi = 0$. The maximum distance between the branches of $\psi = 0$ is $t = 2y_{\max}$ and y will be a maximum when $\theta = \pi$

44 SOLVING PROBLEMS IN FLUID MECHANICS

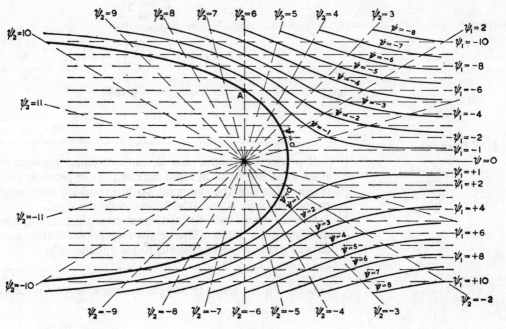

Figure   4.8

so that
$$t = 2 \times \frac{m\pi}{2\pi U} = \frac{m}{U}$$

The pressure coefficient C_p is the pressure increase at any point divided by the reference pressure. If the $\psi = 0$ streamline cuts the y-axis at A and V_A and p_A are the pressures at this point while the pressure is p_0 and velocity U in the undisturbed stream, by Bernoulli's equation:

$$p_A + \tfrac{1}{2}\rho V_A{}^2 = p_0 + \tfrac{1}{2}\rho U^2$$

Excess pressure at A $= p_A - p_0 = \tfrac{1}{2}\rho(U^2 - V^2)$

The reference pressure is taken as the velocity pressure $\tfrac{1}{2}\rho U^2$ so that

$$C_p = 1 - \frac{V^2}{U^2}$$

At the point A $\qquad \theta = \pi/2$

and $\qquad\qquad y_A = \dfrac{m\pi/2}{2\pi U} = \dfrac{m}{4U}$

The velocity V is the resultant of the velocity of the uniform flow $-U$ parallel to the x-axis and the velocity of the flow due to the source which, at A, will be parallel to the y-axis and equal to

$$\frac{m}{2\pi y_A} = \frac{2U}{\pi}$$

$$V^2 = (-U)^2 + \left(\frac{2U}{\pi}\right)^2 = 1{\cdot}405U^2$$

and
$$C_p = 1 - \frac{V^2}{U^2} = -0{\cdot}405$$

4.7 Doublet

> If a source A and a sink B, both of equal strength m, are situated on the x-axis at distances s and $-s$ respectively from the origin, show that the streamlines of the resultant flow are circles passing through A and B with centres on the y-axis.
>
> What is meant by a doublet? Show that the combination of a doublet with a uniform stream gives the potential flow around a circular cylinder corresponding to the streamline $\psi = 0$.

Solution. Figure 4.9(a) shows the arrangement of source and sink

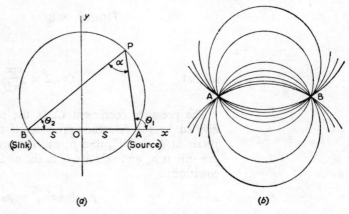

Figure 4.9

Let P be any point and θ_1 and θ_2 the angles made by AP and BP with the x-axis. Combining the flows the stream function for P is given by

$$\psi = \frac{m\theta_1}{2\pi} - \frac{m\theta_2}{2\pi} = \frac{m}{2\pi}(\theta_1 - \theta_2) = \frac{m\alpha}{2\pi}$$

For any given streamline ψ is constant, therefore α is constant and the streamlines are therefore circles passing through A and B, as shown in Fig. 4.9(b). To find the equation for ψ,

$$\tan \theta_1 = \frac{y}{x - s} \qquad \tan \theta_2 = \frac{y}{x + s}$$

$$\tan \alpha = \tan (\theta_1 - \theta_2) = \frac{2ys}{x^2 + y^2 - s^2}$$

$$\alpha = \tan^{-1} \frac{2ys}{x^2 + y^2 - s^2}$$

and
$$\psi = \frac{m}{2\pi} \tan^{-1} \frac{2ys}{x^2 + y^2 - s^2}$$

If the source and sink are allowed to approach until $2s$ is very small but finite and the product $2ms = \mu$ is kept constant the result is a *doublet*, which is an infinite source, and sink an infinitesimal distance apart. The strength μ of a doublet is measured in $m^2/s \times m = m^3/s$. As s tends to zero and the tangent of the angle equals the angle in radians for a small angle, the stream function of the doublet becomes

$$\psi = \frac{\mu}{4\pi} \frac{2y}{(x^2 + y^2)}$$

The streamlines are circles tangential to the x-axis with centres on the y-axis. When $x = 0$

$$\psi = \frac{\mu y}{2\pi y^2}, \quad y = 0 \quad \text{or} \quad \frac{\mu}{2\pi\psi}$$

Figure 4.10 shows the combination of a doublet with a uniform flow. The streamline $\psi = 0$ is a circle and can be replaced by a solid cylinder. The streamlines outside the circle then represent the flow round a cylinder.

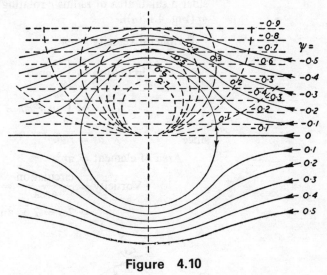

Figure 4.10

4.8 Circulation and vorticity

Define (*a*) circulation, (*b*) vorticity and show that a free vortex in which the velocity is inversely proportional to the radius the vorticity is zero.

Solution. (*a*) The circulation around any closed path PQ in the fluid is defined as the integral taken round the path of the component of the velocity along the path. Let the velocity of the fluid at P be V, Fig.

4.11(a), and consider a small length ds of the closed path which makes an angle θ to the streamline at P, then velocity in direction of $ds = V \cos \theta$ and

$$\text{Circulation } K = \oint V \cos \theta \, ds$$

where $\oint$ means the line integral around the closed path.

If the area enclosed is subdivided at RS, Fig. 4.11(b), the circulations along RS are equal and opposite for the two parts so that

Circulation round PRQS = circulation round PRS

+ circulation round RQS

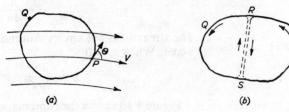

(a)   (b)

Figure 4.11

(b) The vorticity G is the circulation per unit area at a point. Consider a small area of radius r rotating with an average angular velocity ω (Fig. 4.12(a)).

Peripheral velocity $= \omega r$

$$\text{Circulation} = \int_0^{2\pi} \omega r \, ds$$

$$= \int_0^{2\pi} \omega r^2 d\theta = 2\pi r^2 \omega$$

since $\quad\quad ds = r d\theta.$

Area of element $= \pi r^2$

$$\text{Vorticity} = \frac{\text{circulation}}{\text{area}} = \frac{2\pi r^2 \omega}{\pi r^2}$$

$$= 2\omega = 2 \times \text{angular velocity}$$

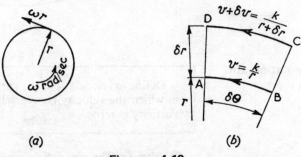

(a)   (b)

Figure 4.12

Fig. 4.12(b) shows a small element at a point in a free vortex.

$$\text{Velocity in AB} = K/r = \text{constant}$$

$$\text{Velocity in CD} = \frac{K}{r + \delta r} = \text{constant}$$

$$\text{Velocity in AD} = \text{velocity in BC} = \text{zero}$$

$$\text{Length of AB} = r\delta\theta$$

$$\text{Length of CD} = (r + \delta r)\delta\theta$$

Circulation round ABCD is the line integral taken anti-clockwise and since the velocities in AB and CD are constant,

$$K = -v \times \text{AB} + (v + \delta v) \times \text{CD}$$

$$= -\frac{K}{r} \times r\delta\theta + \frac{K}{r + \delta r} \times (r + \delta r)\delta\theta$$

$$= 0$$

4.9 Free vortex

Show that in polar co-ordinates the relations between the stream function ψ and the radial component u' and tangential component v' of the velocity at a point whose co-ordinates are (r, θ) are

$$u' = +\frac{\partial \psi}{r\partial \theta} \quad \text{and} \quad v' = -\frac{\partial \psi}{\partial r}$$

Hence show that the stream function for a free vortex of circulation K is

$$\psi = -\frac{K}{2\pi} \log_e\left(\frac{r}{a}\right)$$

where a is the radius of the streamline $\psi = 0$.

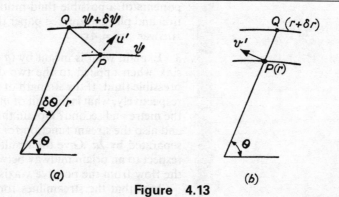

Figure 4.13

Solution. In Fig. 4.13(*a*), P and Q are points on the streamlines ψ and $\psi + \delta\psi$ at radius r from the origin O. Arc PQ subtends an angle $\delta\theta$ at O.

$$\delta\psi = \text{flow across PQ} = u' \times r\delta\theta$$

In the limit

$$u' = + \frac{\partial \psi}{r \partial \theta}$$

In Fig. 4.13(b), P and Q are points on the streamlines ψ and $\psi + \delta \psi$ on the same radius at distances r and $r + \delta r$ from O.

$$\delta \psi = \text{flow across PQ (clockwise about O)}$$
$$= -v' \delta r$$

In the limit, $\quad v' = - \dfrac{\partial \psi}{\partial r}$

For a free vortex, circulation $K = 2\pi r v'$

$$v' = \frac{K}{2\pi r} = - \frac{\partial \psi}{\partial r}$$

Integrating from $r = a$ when $\psi = 0$ to $r = r$ when $\psi = \psi$

$$\psi = - \frac{K}{2\pi} \log_e \left(\frac{r}{a} \right)$$

Problems

1 Obtain the stream functions for (a) a uniform flow with a velocity of 5 m/s parallel to the positive direction of the x-axis; (b) a uniform flow with a velocity of 10 m/s parallel to the positive direction of the y-axis; (c) the flow resulting from the combination of (a) and (b).
Answer $\quad \psi_1 = 5y, \ \psi_2 = -10x, \ \psi_3 = 5y - 10x$

2 Given the stream function for a two-dimensional field of flow, show how to obtain the x and y velocity components at a point in the field.

If $u = 2x, v = -2y$ are respectively the x and y velocity components of a possible fluid motion, determine the stream function and plot on squared paper the streamlines $\psi = 1, 2$ and 3.
Answer $\quad \psi = 4xy$

3 Explain what is meant by (a) a point source and (b) a point sink when applied to the two-dimensional flow of an incompressible fluid. If the strength of these flow forms are m and $-m$ respectively, what is the unit of m when the fundamental units are the metre and second? Obtain the stream function for both cases and also the stream function for a combination of the two when separated by $2a$. Give the result in Cartesian co-ordinates with respect to an origin midway between source and sink, measuring the flow from the positive x-axis which is the line joining them.

Show that the streamlines for the combined flow are circles and draw the streamline $\psi = 3 \cdot 5$ for the case where $m = 20$ and $a = 10$ and indicate the direction of flow by means of arrows.
Answer $\quad \dfrac{m\theta}{2\pi}, \ -\dfrac{m\theta}{2\pi}, \ \dfrac{m}{2\pi} \tan^{-1} \dfrac{2ya}{x^2 + y^2 - a^2}$

4 Considering the two-dimensional flow of an incompressible fluid obtain the stream function ψ for (a) linear flow having uniform velocity U parallel to the x-axis and (b) a source having strength m.

Obtain the streamlines $\psi = 0$ and $\psi = \pm 10$ for the flow resulting from the combination of linear flow having $U = -20$ m/s with a source whose centre lies on the x-axis and for which $m = 15$ m²/s.

Draw the streamlines for the linear flow and source to show values of ψ increasing by intervals of 1 m²/s, using a scale of 1 cm = 4 m²/s for the linear flow streamlines, which need be drawn only for values of ψ between the limits ± 16.

What is the special significance of the streamline $\psi = 0$?

5 A point source and a point sink of equal strength are separated by a distance $2a$. Derive the equations for the streamlines of a potential flow between them for the limiting case when $a = 0$. Hence show how this flow may be combined with a uniform indefinite flow to determine the nature of the flow around a solid cylinder and that the equation to the streamlines is given by

$$\psi = -Vy\left(1 - \frac{a^2}{x^2 + y^2}\right)$$

where $-V$ is the velocity of the uniform flow in the direction of $y = 0$, a the radius of the cylinder and x and y co-ordinates with origin at the centre of the cylinder.

Explain briefly the difference between this ideal flow and an actual flow with similar boundaries.

6 Define a "doublet".

Considering two-dimensional flow of an incompressible fluid, show that the stream function for a doublet whose axis is horizontal and strength is μ is given by

$$\psi = -(\mu/2\pi)\,\frac{\sin\theta}{r}$$

Draw full size the streamline for $\psi = \pm 20$ for a doublet whose strength is 5π units.

7 Explain the meaning of the term "stream function". Obtain the stream function for (a) a simple source and a sink of strength m at a distance $2s$ apart, (b) a doublet of strength μ. What flow does the combination of a doublet with a uniform stream represent?

Answer $\psi = \dfrac{m}{2\pi}\,\tan^{-1}\dfrac{2ys}{x^2 + y^2 - s^2}$, $\psi = \dfrac{\mu}{4\pi}\cdot\dfrac{2y}{x^2 + y^2}$

8 A source of strength m is placed on the x-axis at a distance s from a wall which is parallel to the y-axis and passes through the origin. Draw the streamlines for the resulting flow and show that it is the same as that produced by combining the original source with an equal source situated on the x-axis at a distance *s behind* the wall.

9 Sketch the streamlines resulting from the combination of a source of strength $m = 24$ units and a free vortex of strength $K - 40$ units. Draw sufficient of the pattern within the vortex streamline $\psi = 6$ to sketch also two equipotential lines. If the potential lines were regarded as streamlines what combination of flows would they represent?

10 A point source of strength 24 m²/s located at the origin of co-ordinates is combined with a uniform flow of velocity 10 m/s parallel to the x-axis in the direction from positive to negative x. Find the distance of the stagnation point from the origin and determine the thickness of the body contour for values of $\theta = \pi/2$ and π. Calculate the velocity at the body contour for $\theta = \pi/2$.

Answer   0·382 m, 1·2 m, 2·4 m, 11·85 m/s

5

Gases at rest

Both gases and liquids are fluids, but a gas is readily compressible while a liquid is not. So far we have assumed that the fluid is incompressible and of constant density. The same principles can be applied to a gas but the working must be modified to allow for the variation of its density with pressure and temperature.

Steady flow for a gas occurs when the mass passing successive cross-sections is constant.

The *continuity of flow equation* for a gas will therefore be

$$\rho A v = \text{constant}$$

where ρ = mass density, A = area of cross-section and v = velocity. If V is the volume per unit mass then for any two sections

$$\rho_1 A_1 v_1 = \rho_2 A_2 v_2 \quad \text{or} \quad \frac{A_1 v_1}{V_1} = \frac{A_2 v_2}{V_2}$$

For a liquid $\rho_1 = \rho_2$ and the continuity equation reduces to the form previously used:

$$A_1 v_1 = A_2 v_2$$

Fundamental equations

A knowledge of the fundamental relationships for gases is necessary. *Boyle's Law*. At a constant temperature T,

$$pV = \text{constant}$$

where p = absolute pressure and V = volume of gas. *Charles' Law*. At constant pressure,

$$\frac{V}{T} = \text{constant}$$

or at constant volume,

$$\frac{p}{T} = \text{constant}$$

where T = absolute temperature.

Characteristic equation. Boyle's law and Charles' law can be combined to give

$$\frac{pV}{T} = \text{constant}$$

For a mass of gas m of density ρ, the volume V is proportional to m, therefore

$$\frac{pV}{T} = mR$$

where R is the gas constant.

For unit mass of the gas this equation becomes

$$p = \rho RT$$

where ρ is the mass density of the gas.

This is the *characteristic equation* or *equation of state* of the gas.

Since $R = pV/mT$ the units of R are those of work per unit mass per unit of temperature. The value of R depends on the system of units. For air,

$$R = 287\,\text{J/kg-K} = 96\,\text{ft-lbf/lb °C} = 53\cdot3\,\text{ft-lbf/lb °F}.$$

Specific heats. A gas has two different specific heats.

(*a*) Specific heat at constant volume C_v. When unit mass of gas has its temperature changed from T_1 to T_2 at constant volume

$$\text{Heat supplied} = H = 1 \times C_v(T_2 - T_1)$$

Since there is no change in volume no external work is done.

$$\text{Increase of internal energy} = C_v(T_2 - T_1)$$

(*b*) Specific heat at constant pressure C_p. When the gas is heated from temperature T_1 to T_2 external work is done.

$$\text{Heat supplied per unit mass} = C_p(T_2 - T_1) \text{ heat units}$$

Only part of this energy is used to heat the gas, the rest going to external work, thus C_p is greater than C_v.

$$C_p(T_2 - T_1) = C_v(T_2 - T_1) + \text{external work in heat units}$$

It can be shown that $R = J(C_p - C_v)$.

Mechanical equivalent of heat J. The relation of mechanical energy to heat energy is

$$\text{Mechanical energy} = J \times \text{energy in heat units}$$

In SI the units of mechanical energy and heat energy are the same so that $J = 1$. In the ft-slug-s system,

$$J = 1400\,\text{ft-lb/Chu} = 778\,\text{ft-lb/Btu}$$

5.1 Work done during compression

Derive expressions for the work done per unit mass, when the gas is compressed (*a*) isothermally, (*b*) according to the law pV^n = constant, (*c*) adiabatically.

Solution. (a) If the compression is carried out isothermally the temperature of the gas is maintained constant.

The characteristic equation $p = \rho RT$ becomes

$$pv = \text{constant}$$

where v = specific volume = volume/unit mass = $1/\rho$. If p_1 and v_1 are the initial absolute pressure and specific volume and p_2 and v_2 the corresponding final values, referring to Fig. 5.1,

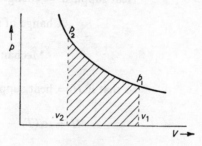

Figure 5.1

Work done/unit mass = area under $p - v$ curve = $\int_{v2}^{v1} p\,dv$

But
$$pv = \text{constant} = p_1 v_1$$

so that
$$p = p_1 v_1 \left(\frac{1}{v}\right)$$

$$\text{Work done} = p_1 v_1 \int_{v1}^{v2} \frac{dv}{v}$$

Work done/unit mass in isothermal compression = $p_1 v_1 \log_e \dfrac{v_1}{v_2}$

$$= RT \log_e \frac{v_1}{v_2}$$

(b) If the law is $pv^n = \text{constant}$

$$pv^n = p_1 v_1{}^n \quad \text{and} \quad p = p_1 v_1{}^n v^{-n}$$

Work done/unit mass = $\displaystyle\int_{v2}^{v1} p.dv$

$$= p_1 v_1{}^n \int_{v2}^{v1} v^{-n} dv$$

$$= \frac{p_1 v_1{}^n}{1 - n}\{v_1^{1-n} - v_2^{1-n}\}$$

$$= \frac{1}{1 - n}\{p_1 v_1 - p_1 v_1{}^n v_2^{1-n}\}$$

or since $p_1 v_1^n = p_2 v_2^n$

$$\text{Work done/unit mass} = \frac{p_1 v_1 - p_2 v_2}{1 - n}$$

$$= \frac{R(T_2 - T_1)}{n - 1}$$

(c) If the compression is carried out adiabatically no heat enters or leaves the system.

For any mode of compression, considering unit mass,

Heat supplied = change of internal energy + work done in heat units

$$\text{Change of internal energy} = c_v(T_2 - T_1)$$

$$\text{Mechanical work done} = \frac{p_2 v_2 - p_1 v_1}{(1 - n)}$$

so that if H = heat supplied

$$H = c_v(T_2 - T_1) + \frac{(p_2 v_2 - p_1 v_1)}{J(1 - n)}$$

Now $\quad R = J(c_p - c_v)\quad$ or $\quad c_v = \dfrac{R}{J\left(\dfrac{c_p}{c_v} - 1\right)}$

Also $R(T_2 - T_1) = (p_2 v_2 - p_1 v_1)$

Thus $\qquad H = \dfrac{p_2 v_2 - p_1 v_1}{J\left(\dfrac{c_p}{c_v} - 1\right)} + \dfrac{p_2 v_2 - p_1 v_1}{J(1 - n)}$

For an adiabatic change $H = 0$ so that

$$n = \frac{c_p}{c_v} = \gamma$$

Thus for an adiabatic change $\quad pv^\gamma = \text{constant}$

and from (b) $\quad$ Work done/unit mass $= \dfrac{p_2 v_2 - p_1 v_1}{(\gamma - 1)}$

$$= \frac{R(T_2 - T_1)}{(\gamma - 1)}$$

5.2 Velocity of a pressure wave

Obtain expressions for the velocity of propagation of a pressure wave through a gas (a) assuming an isothermal process, (b) assuming an adiabatic process.

Calculate the velocity of sound in air at a pressure of 689 kN/m^2 abs and a temperature of 20°C. R for air = 287 J/kg-K, $\gamma = 1\cdot4$. Assume an adiabatic process.

Solution. In example 10.5 (p. 146) it will be shown that the velocity of propagation of a pressure wave in an elastic fluid of bulk modulus K is given by $a = \sqrt{(K/\rho)}$.

By definition, $\quad$ Bulk modulus $= \dfrac{\text{pressure stress intensity}}{\text{volumetric strain}}$

or $$K = -V\frac{dp}{dV}$$

(a) For an isothermal process, considering unit mass of volume v

$$pv = \text{constant}$$
$$pdv + vdp = 0$$
$$p = -v\,\frac{dp}{dv} = K$$

Velocity of propagation $a = \sqrt{\dfrac{p}{\rho}}$

(b) For an adiabatic process $pv^{\gamma} = \text{constant}$
$$\gamma pv^{\gamma-1}dv + v^{\gamma}dp = 0$$
$$\gamma p = -v\,\frac{dp}{dv} = K$$

Velocity of propagation $a = \sqrt{\dfrac{\gamma p}{\rho}}$

but $$\frac{p}{\rho} = RT$$

Therefore $$a = \sqrt{(\gamma RT)}$$
Putting $\gamma = 1{\cdot}4$, $\quad R = 287\,\text{J/kg-K}$, $\quad T = 20°\text{C} = 293\,\text{K}$
$$v = \sqrt{(1{\cdot}4 \times 287 \times 293)} = \textbf{343\,m/s}$$

5.3 Variation of atmospheric pressure with altitude

Derive expressions for the variation of pressure with altitude in the atmosphere assuming (a) a constant temperature drop t per unit increase of altitude, (b) an isothermal condition, (c) an adiabatic condition.

Solution. Consider a vertical column of air of area A (Fig. 5.2) and let the pressure at altitude h be p and at altitude $h + \delta h$ be $p + \delta p$. If w is the specific weight of the air at altitude h, for vertical equilibrium of the air between h and $h + \delta h$,

Upward force due to pressure difference

$$= \text{weight of air between sections}$$
$$-A\delta p = \rho g A \delta h$$
$$\delta h = -\frac{\delta p}{\rho g}$$

where $\rho = $ mass density at pressure p.

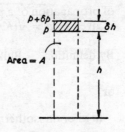

Figure 5.2

But
$$p = \rho RT \quad \text{or} \quad \rho = p/RT$$

so that
$$\delta h = - \frac{RT}{g} \frac{\delta p}{p} \tag{1}$$

(a) If $\quad T_0 =$ absolute temperature at ground level

$p_0 =$ absolute pressure at ground level

$T =$ absolute temperature at altitude h

$\quad = T_0 - th$

from equation (1)

$$\delta h = - \frac{R}{g} (T_0 - th) \frac{\delta p}{p}$$

$$\frac{dp}{p} = - \frac{gdh}{R(T_0 - th)}$$

Integrating between limits of p_0 and p_1, and 0 and h_1

$$\log_e \frac{p_1}{p_0} = \frac{g}{Rt} \log_e \left(\frac{T_0 - th_1}{T_0} \right)$$

$$= \frac{g}{Rt} \log_e \left(1 - \frac{th_1}{T_0} \right) \tag{2}$$

(b) For isothermal conditions T is a constant in equation (1) which can therefore be integrated immediately.

$$\int_0^{h_1} dh = - \frac{RT}{g} \int_{p_0}^{p_1} \frac{dp}{p}$$

$$\log_e \frac{p_1}{p_0} = - \frac{gh_1}{RT}$$

$$\frac{p_1}{p_0} = e^{-gh_1/RT} \tag{3}$$

(c) For adiabatic conditions
$$pv^\gamma = p_1 v_1^\gamma$$

or since $pv = RT$

$$p = p_1 \left(\frac{T_1}{T} \right)^{\gamma/(1-\gamma)}$$

so that

$$dp = -\left(\frac{\gamma}{1-\gamma}\right) p_1 T_1^{\gamma/(1-\gamma)} T^{-1/(1-\gamma)} dT$$

Substituting these values of p and dp in equation (1)

$$dh = \frac{-RT\left\{-\left(\dfrac{\gamma}{1-\gamma}\right) p_1 T_1^{\gamma/(1-\gamma)} T^{-1/(1-\gamma)} dT\right\}}{g \times p_1 T_1^{\gamma/(1-\gamma)} T^{-\gamma/(1-\gamma)}}$$

$$dh = \left(\frac{\gamma}{1-\gamma}\right) \frac{RdT}{g}$$

Temperature gradient $= \dfrac{dT}{dh} = -\left(\dfrac{\gamma-1}{\gamma}\right) \dfrac{g}{R}$ \hfill (4)

Integrating $\displaystyle \int_0^{h_1} dh = \left(\frac{\gamma}{1-\gamma}\right) \frac{R}{g} \int_{T_0}^{T_1} dT$

$$h_1 = \left(\frac{\gamma}{1-\gamma}\right) \frac{RT_0}{g} \left(\frac{T_0}{T_1} - 1\right)$$

or since for adiabatic conditions

$$\frac{T_1}{T_0} = \left(\frac{p_1}{p_0}\right)^{(\gamma-1)/\gamma}$$

then

$$h_1 = \left(\frac{\gamma}{1-\gamma}\right) \frac{RT_0}{g} \left\{\left(\frac{p_1}{p_0}\right)^{(\gamma-1)/\gamma} - 1\right\}$$

5.4 Variation of atmospheric pressure assuming uniform temperature lapse rate

Find the pressure and density of the atmosphere at a height of 750 m when the pressure and temperature at sea level are 762 mm of mercury and 17°C assuming a uniform decrease of temperature with altitude at the rate of 4·92°C per 1000 m. R for air is 287 J/kg-K.

Solution. From example **5.3**(a), equation (2),

$$\log_e \frac{p_1}{p_0} = \frac{g}{Rt} \log_e \left(1 - \frac{th_1}{T_0}\right)$$

$$p_1 = p_0 \left(1 - \frac{th_1}{T_0}\right)^{g/Rt}$$

$$p_0 = \frac{762}{1000} \times 13\cdot6 \times 9\cdot81 \times 10^3 = 101\cdot66 \times 10^3\,\text{N/m}^2$$

$$t = 0\cdot00492°\text{C/m}, \quad h_1 = 750\,\text{m}, \quad T_0 = 290\,\text{K}$$

$$p_1 = 101\cdot66 \times 10^3 \left(1 - \frac{0\cdot00492 \times 750}{290}\right)^{9\cdot81/287 \times 0\cdot00492}$$

$$= 101\cdot66 \times 10^3 \times 0\cdot98728^{6\cdot95} = 93 \times 10^3\,\text{N/m}^2$$

$$= \textbf{697 mm of mercury}$$

$$\frac{p_1}{\rho_1} = RT_1 \qquad \rho_1 = \frac{p_1}{RT_1} = \frac{p_1}{R(T_0 - th_1)}$$

Density at $2500\,\text{ft}$

$$= \rho_1 = \frac{93 \times 10^3}{287(290 - 0\cdot00492 \times 750)}$$

$$= 1\cdot132\,\text{kg/m}^3$$

5.5 Variation assuming $p = k\rho^n$

Atmospheric conditions are such that the relation between pressure p and density ρ as altitude increases is given by $p = k\rho^n$ where k and n are constants. Show that the relation between temperature T and altitude H is given by

$$gH = \frac{nR}{(n-1)}\,(T_0 - T)$$

where T_0 is the temperature at zero altitude and R is the gas constant.

At zero altitude the pressure is $101\,\text{kN/m}^2$, the temperature is $15°\text{C}$ and the air density is $1\cdot225\,\text{kg/m}^3$. Find (a) the temperature, and (b) the density, at an altitude of 3000 m, assuming $n = 1\cdot25$.

Solution. Example **5.3**(*c*) gives the temperature gradient for conditions following the law $pv^\gamma = $ constant or, putting $v = 1/\rho g$, the law $p = k\rho^\gamma$. Replacing γ by n the solution is applicable to conditions following the law $p = k\rho^n$ and equation (4) becomes

$$\frac{dT}{dh} = -\left(\frac{n-1}{n}\right)\frac{g}{R}$$

Integrating between the limits of $T = T_0$ when $h = 0$ and $T = T$ when $h = H$

$$\int_0^H dh = -\left(\frac{n}{n-1}\right)\frac{R}{g}\int_{T_0}^T dT$$

$$gH = \frac{nR}{n-1}\,(T_0 - T) \tag{1}$$

(a) From equation (1)

$$T_0 - T = \frac{gH(n-1)}{nR}$$

$$T = T_0 - \frac{gH(n-1)}{nR}$$

or since $p_0 = \rho_0 R T_0$, $R = \dfrac{p_0}{\rho_0 T_0}$

$$T = T_0 - \frac{gH(n-1)\rho_0 T_0}{np_0}$$

Putting $\quad H = 3000\,\text{m}, \quad n = 1\cdot25$

$$T_0 = (273 + 15) = 288\,\text{K}$$

$$p_0 = 101 \times 10^3\,\text{N/m}^2, \quad \rho_0 = 1\cdot225\,\text{kg/m}^3$$

$$T = 288 - \frac{9\cdot81 \times 3000 \times 0\cdot25 \times 1\cdot225 \times 288}{1\cdot25 \times 101 \times 10^3}$$

$$= 288 - 20\cdot56 = 267\cdot44\,\text{K}$$

$$T = -5\cdot56°\text{C}$$

(b) From the characteristic equation

$$p = \rho R T \quad \text{or} \quad \rho = \frac{p}{RT}$$

Similarly

$$\rho_0 = \frac{p_0}{RT_0}$$

$$\frac{\rho}{\rho_0} = \frac{p}{p_0}\frac{T_0}{T} \quad \text{or since} \quad \frac{p}{p_0} = \left(\frac{\rho}{\rho_0}\right)^n$$

$$\frac{\rho}{\rho_0} = \left(\frac{\rho}{\rho_0}\right)^n \frac{T_0}{T}, \qquad \frac{\rho}{\rho_0} = \left(\frac{T}{T_0}\right)^{1/(n-1)}$$

$$\rho = \rho_0 \left(\frac{T}{T_0}\right)^{1/(n-1)}$$

$$= 1\cdot225 \times \left(\frac{267\cdot44}{288}\right)^{1/0\cdot25}$$

$$= 1\cdot225 \times 0\cdot9286^4 = \mathbf{0\cdot911\,kg/m^3}$$

5.6 Standard atmosphere

The temperature lapse rate in the I.C.A.N. standard atmosphere may be taken as $6\cdot5°\text{C}$ per 1000 m. Show that this corresponds to the law

$$\frac{p}{p_0} = \left(\frac{\rho}{\rho_0}\right)^{1\cdot235}$$

Also find the pressure at a height of 3000 m in this atmosphere taking datum conditions as $15°\text{C}$ and $101\,\text{kN/m}^2$. For air $R = 287\,\text{J/kg-K}$ units.

Solution. Let the law be $\dfrac{p}{p_0} = \left(\dfrac{\rho}{\rho_0}\right)^n$

From the characteristic equation

$$\frac{p}{p_0} = \frac{\rho T}{\rho_0 T_0} = \left(\frac{p}{p_0}\right)^{1/n}\frac{T}{T_0}$$

$$\frac{T}{T_0} = \left(\frac{p}{p_0}\right)^{(n-1)/n}$$

Both T and p vary with altitude h.

Differentiating with respect to h

$$\frac{1}{T_0}\frac{dT}{dh} = \frac{n-1}{n p_0{}^{(n-1)/n}} p^{-1/n}\frac{dp}{dh}$$

For atmospheric equilibrium from example **5.3**, equation (1),

$$\frac{dp}{dh} = -\frac{gp}{RT}$$

so that

$$\frac{dT}{dh} = -\frac{n-1}{n} T_0 \frac{p^{-1/n}}{p_0{}^{(n-1)/n}}\frac{gp}{RT}$$

$$= -\frac{(n-1)}{n}\frac{T_0}{T}\left(\frac{p}{p_0}\right)^{(n-1)/n}\frac{g}{R} = -\frac{(n-1)}{n}\frac{g}{R}$$

For a lapse rate of $6\cdot5°C$ per $1000\,m$

$$\frac{dT}{dh} = -\frac{6\cdot5}{1000} = -\frac{n-1}{n}\times\frac{g}{287}$$

$$6\cdot5n = \frac{1000\times9\cdot81}{287}(n-1)$$

$$n = 1\cdot235$$

and

$$\frac{p}{p_0} = \left(\frac{\rho}{\rho_0}\right)^{1\cdot235}$$

At a height of $3000\,m$, $T = 288 - 19\cdot5 = 268\cdot5\,K$

$$p = p_0\left(\frac{T}{T_0}\right)^{n/n-1} = 101\times\left(\frac{268\cdot5}{288}\right)^{1\cdot235/0\cdot235}$$

$$= 69\cdot9\,kN/m^2$$

Problems

1 A quantity of gas is compressed isothermally from initial conditions of $0\cdot6\ m^3$ and $101\ kN/m^2$ absolute to a final pressure of $707\ kN/m^2$ absolute. Find the work done on the gas.
Answer $118\ kN\text{-}m$

2 A quantity of gas weighing $0\cdot57\ kg$ expands from a pressure of $828\ kN/m^2$ abs to a pressure of $103\ kN/m^2$ abs in such a way that the gas does not receive or reject any heat. The initial temperature is $200°C$. Find the work done, the final temperature and the change in internal energy if $c_p = 1\cdot00\ kJ/kg\text{-}°C$ and $c_v = 0\cdot716\ kJ/kg\text{-}K$.
Answer $86\cdot5\ kJ$, $261\ K$, $86\cdot5\ kJ$

3 Show that a disturbance is propagated with velocity $\sqrt{(K/\rho)}$ in a fluid of density ρ and bulk modulus of elasticity K. Calculate the velocity of a pressure wave transmitted through a liquid hav-

ing a specific gravity of 0·85 and a bulk modulus of 1·958 GN/m².
Answer 1518 m/s

4 Calculate the velocity of sound in air having a pressure of 64·7 kN/m² abs and a temperature of $-9°C$, (*a*) assuming an isothermal process, (*b*) assuming an adiabatic process. R for air $= 287$ J/kg-K and $\gamma = 1·4$.
Answer 275 m/s, 325 m/s

5 An air vessel of 56 dm³ capacity was pumped up with air and at the end of the pumping operation it showed a pressure of 7650 kN/m² and a temperature of 44°C. The air then cooled to atmospheric temperature 15°C, after which leakage occurred down to 2060 kN/m², when the leak was stopped, the temperature of the air then being 3°C. Find (*a*) how much heat was lost or gained during leakage by the air remaining in the vessel, assuming the index of expansion constant. $R = 287$ J/kg-K, $c_v = 0·71$ kJ/kg-K.
Answer 97 kJ, + 122·8 kJ

6 Calculate the velocity of sound in air having a pressure of 10 bar abs and a temperature of 20°C, (*a*) assuming an isothermal process, (*b*) assuming an adiabatic process, $\gamma = 1·4$, $R = 287$ J/kg-K.
Answer 290 m/s, 343 m/s

7 Deduce an expression for the variation of barometric pressure with altitude assuming that the temperature remains constant.
At sea level the barometer stands at 742 mm. What will be the height of the barometer at an altitude of 1350 m if the temperature remains constant at 15°C? R for air is 287 J/kg-K.
Answer 631 mm mercury

8 Find the pressure and density of the atmosphere at a height of 900 m when the pressure and temperature at ground level are 760 mm of mercury and 15°C. Assume that the atmospheric temperature diminishes with height at a uniform rate of 5 deg C. per 1000 m. For air $R = 287$ J/kg-K.
Answer 683 mm mercury, 1·118 kg/m³

9 Given that the barometric pressure is ρ_0 at ground level where the temperature is 15°C, prove that the pressure ρ at a height hm is given by the expression

$$\log(p/p_0) = A \log (1 - Bh)$$

in which A and B are constants.
If the temperature of a quiescent atmosphere diminishes with height at a uniform rate of 6·5°C per 1000 m and for air $R = 287$ J/kg-K, find the values of A and B.
Answer 0·536, 2·25 × 10⁻⁵

10 In the standard atmosphere the temperature at sea level is taken at 15°C, and it is further assumed that the lapse rate is constant at 6·5°C per 1000 m until the stratosphere is reached in which the temperature remains constant at $-56·5°C$.

Calculate the relative pressure and relative density of the atmosphere at a height of 13 700 m taking the sea-level values as unity and the gas constant $R = 287$ J/kg-K.

Answer 0·146, 0·194

11 If the pressure and temperature of the atmosphere at ground level are 101 kN/m² and 15°C respectively, calculate the pressure and density at an altitude of 4870 m assuming an adiabatic atmosphere. Find also the mean temperature gradient up to this altitude. $R = 287$ J/kg-K and $\gamma = 1\cdot4$.

Answer 53·73 kN/m², 0·78 kg/m³, 9·86°C per 1000 m

12 Find the pressure and density of the atmosphere at an altitude of 3900 m assuming an isothermal atmosphere. At sea level $p = 101$ kN/m² and $t = 15°C$, R for air = 287 J/kg-K.

Answer 63·58 kN/m²

13 If the pressure and temperature of the atmosphere at ground level are 101 kN/m² and 10°C respectively, calculate the pressure and density at an altitude of 6100 m assuming an adiabatic atmosphere. Find also the mean temperature gradient up to this altitude. $R = 287$ J/kg-K and $\gamma = 1\cdot4$.

Answer 46·05 kN/m², 0·72 kg/m³, 9·77°C per 1000 m

14 Show that if the air temperature is assumed not to vary with altitude, the barometric pressure at altitude H is given by $p_H = p_0 e^{-gH/RT}$ where p_0 is the barometric pressure at ground level and R and T are the gas constant and absolute temperature.

Hence calculate the pressure and density of air at an altitude of 3050 m if the temperature is 17°C and the barometric pressure at ground level is 101 kN/m². $R = 287$ J/kg-K.

Answer 70·5 kN/m², 0·848 kg/m³

15 Explain the conditions necessary for a state of convective equilibrium to occur in the atmosphere.

Provided that these conditions occur and assuming that the ratio of the specific heats for air does not change with increase of altitude, show that the temperature lapse rate is constant.

If the temperature is 15°C and the pressure is 101 kN/m² at zero altitude, calculate: (*a*) the temperature, (*b*) the density of the atmosphere at an altitude of 1200 m.

Take $\gamma = 1\cdot41$ and $R = 287$ J/kg-K.

Answer 3·1°C, 1·102 kg/m³

16 The pressure, temperature and density of the atmosphere at sea level are 101 kN/m² absolute, 15°C, and 1·235 kg/m³ respectively. Assuming that the temperature of the atmosphere diminishes with height at a rate of 0·00656°C/m, find the pressure and density at a height of 7600 m. For air $R = 287$ J/kg-K.

Answer 37·5 kN/m², 0·549 kg/m³

17 Show that the ratio of the atmospheric pressure at an altitude h_1 to that at sea level may be expressed as $p/p_0 = (T/T_0)^n$, a uniform temperature lapse rate being assumed. Find the ratio of the pressures and the densities at 10 700 m and sea

level taking the standard atmosphere as 15°C and a lapse rate of 6·5°C per 1000 m to a minimum of −56·5°C.
Answer 0·237, 0·312

18 Observations taken at ground level show the temperature and pressure of the atmosphere at a given time to be 13°C and 750 mm mercury respectively. Calculate the barometric pressure and the density of the air at a height of 4250 m above the station if the temperature decreases uniformly at a rate of 6·4°C per 1000 m of height. Prove any formula used. $R = 287$ J/kg-K.
Answer 58·68 kN/m², 0·79 kg/m³

19 Show that the ratios of the pressures (p_2/p_1) and densities (ρ_2/ρ_1) for altitudes h_2 and h_1 in an isothermal atmosphere are each given by

$$\frac{p_2}{p_1} = \frac{\rho_2}{\rho_1} = e^{-g(h_2-h_1)/RT}$$

What increase in altitude is necessary in the stratosphere to halve the pressure? Assume a constant temperature of −56·5°C and $R = 287$ J/kg-K.
Answer 4390 m

6

Flow of gases

When a gas flows there will be changes of density and temperature from point to point as well as changes of pressure, velocity and elevation. The continuity of flow equation must be on a mass basis and the discharge measured as the mass flowing per unit time.

Bernoulli's equation must also be modified to take into account the variation of density with pressure.

This chapter covers flow through venturi meters, orifices and nozzles, and an elementary treatment of pipe flow. A comparison with the treatment of these subjects in previous chapters will show that the procedure adopted for a gas is fundamentally the same except for the modification of the continuity equation and Bernoulli's equation to allow for variation of density with pressure.

6.1 Bernoulli's equation

> Derive Bernoulli's equation for the steady frictionless flow of a gas assuming (a) isothermal conditions, (b) adiabatic conditions.

Solution. The derivation of Bernoulli's equation for a gas follows precisely the proof given in example 5.6 (Volume 1) up to equation (1)

$$\delta z + \frac{v\delta v}{g} + \frac{\delta p}{w} = 0$$

from which

$$g\delta z + v\delta v + \frac{\delta p}{\rho} = 0 \tag{1}$$

Integrating along the stream

$$gz + \tfrac{1}{2}v^2 + \int \frac{dp}{\rho} = \text{constant} \tag{2}$$

The term $\int \frac{dp}{\rho}$ cannot be integrated until the relation between p and ρ is known.

(a) Isothermal conditions, $p/\rho = k = RT$ where k is a constant since T is constant.

Substituting in equation (2)

$$gz + \tfrac{1}{2}v^2 + k \int \frac{dp}{p} = \text{constant}$$

$$gz + \tfrac{1}{2}v^2 + k \log_e p = \text{constant} \tag{3}$$

The potential energy z can often be ignored. Considering any two points on a streamline equation (3) becomes

$$k (\log_e p_2 - \log_e p_1) = \frac{v_1{}^2 - v_2{}^2}{2}$$

or

$$RT \log_e \frac{p_2}{p_1} = \tfrac{1}{2}(v_1{}^2 - v_2{}^2)$$

(b) Adiabatic conditions $\dfrac{p}{\rho^\gamma} = k = \text{constant}$

$$\rho = \left(\frac{p}{k}\right)^{1/\gamma}$$

Substituting in equation (2)

$$gz + \tfrac{1}{2}v^2 + k^{1/\gamma} \int p^{-1/\gamma} \, dp = \text{constant}$$

$$gz + \tfrac{1}{2}v^2 + k^{1/\gamma} \frac{\gamma}{\gamma - 1} \cdot p^{(\gamma-1)/\gamma} = \text{constant}$$

or since $k = p/\rho^\gamma$

$$gz + \tfrac{1}{2}v^2 + \left(\frac{\gamma}{\gamma - 1}\right) \frac{p}{\rho} = \text{constant} \qquad (4)$$

For any two points on a streamline, ignoring z,

$$\left(\frac{\gamma}{\gamma - 1}\right)\left(\frac{p_2}{\rho_2} - \frac{p_1}{\rho_1}\right) = \tfrac{1}{2}(v_1{}^2 - v_2{}^2)$$

6.2 Steady flow energy equation

> (a) Derive from first principles the steady flow energy equation for a fluid.
>
> (b) The turbine of a jet engine is supplied with a steady flow of gas at a temperature of 866°C and absolute pressure 717 kN/m² and velocity 162 m/s which is discharged at 624°C, abs pressure 217 kN/m² and a velocity of 300 m/s. If the process is adiabatic what is the work done per unit mass of gas?

Solution. (a) The steady flow energy equation can be derived from the first law of thermodynamics which states:

For any mass system, the net heat energy supplied to the system equals the increase of energy of the system plus the energy leaving the system as work done.

If ΔE is the increase in energy of the system, ΔQ the energy supplied to the system and ΔW the energy leaving the system in a given interval,

$$\Delta E = \Delta Q - \Delta W$$

The energy contained in the system will be in the following forms:

(i) internal energy of the molecules and atoms,
(ii) kinetic energy resulting from the motion of the mass concerned,
(iii) potential energy due to the mass concerned being elevated above datum level and acted upon by gravitational force.

Considering the flow of a fluid through the stream tube shown in Fig. 6.1, fluid enters at section 1, area A_1, with pressure p_1, velocity v_1,

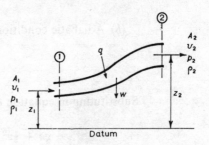

Figure 6.1

density ρ_1 at an elevation z_1 with internal energy per unit mass u_1. The fluid leaves at section 2 where the corresponding values are A_2, p_2, v_2, ρ_2, z_2 and u_2. Between sections 1 and 2 heat energy q per unit mass enters and mechanical energy w per unit mass leaves the system.

For steady flow the mass of fluid per unit time $\dot{m}$ passing sections 1 and 2 will be the same.

Energy of fluid entering in unit time at section 1 is given by

$$E_1 = \text{kinetic energy} + \text{potential energy} + \text{internal energy}$$
$$= \dot{m}(\tfrac{1}{2}v_1{}^2 + gz_1 + u_1)$$

Energy of fluid leaving in unit time at section 2,

$$E_2 = \dot{m}(\tfrac{1}{2}v_2{}^2 + gz_2 + u_2)$$

Change of energy of fluid in unit time,

$$\Delta E = E_2 - E_1 = \dot{m}\{\tfrac{1}{2}(v_2{}^2 - v_1{}^2) + g(z_2 - z_1) + (u_2 - u_1)\}$$

Work done by fluid entering at section 1 is

$$\text{Pressure at section 1} \times \text{volume entering per unit time} = p_1\frac{\dot{m}}{\rho_1}$$

Work done by fluid leaving at section 2 is

$$p_2\frac{\dot{m}}{\rho_2}$$

Energy leaving as work between section 1 and 2 is $\dot{m}w$. Total energy leaving the system as work in unit time,

$$\Delta W = \dot{m}\left(w + \frac{p_2}{\rho_2} - \frac{p_1}{\rho_1}\right)$$

Total energy entering as heat $= \Delta Q = \dot{m}q$.

Then, since $\Delta E = \Delta Q - \Delta W$,

$$\dot{m}\{\tfrac{1}{2}(v_2{}^2 - v_1{}^2) + g(z_2 - z_1) + (u_2 - u_1)\} = \dot{m}q - \dot{m}\left(w + \frac{p_2}{\rho_2} - \frac{p_1}{\rho_1}\right)$$

or for unit mass

$$gz_1 + \tfrac{1}{2}v_1{}^2 + \left(\frac{p_1}{\rho_1} + u_1\right) + q - w = gz_2 + \tfrac{1}{2}v_2{}^2 + \left(\frac{p_2}{\rho_2} + u_2\right)$$

The terms $\left(\dfrac{p_1}{\rho_1} + u_1\right)$ and $\left(\dfrac{p_2}{\rho_2} + u_2\right)$ can be replaced by the enthalpies h_1 and h_2 giving

$$gz_1 + \tfrac{1}{2}v_1{}^2 + h_1 + q - w = gz_2 + \tfrac{1}{2}v_2{}^2 + h_2 \tag{1}$$

which may be written

$$q - w = \Delta(h + \tfrac{1}{2}v^2 + gz) \tag{2}$$

where Δ means "the increase in".

The steady flow energy equation (1) or (2) applies to all fluids (real or ideal), whether liquids, vapours or gases. The assumptions are:

(1) Flow is continuous and steady so that the mass entering section 1 equals the mass leaving section 2 and does not vary with time.

(2) Heat and work are transferred to or from the fluid at a constant rate.

(3) Conditions at any point do not vary with time.

(4) All quantities are constant over the inlet and outlet sections.

(5) Mechanical and thermal energy alone are considered, additional terms would be required for other forms of energy such as electrical energy.

(b) Applying the steady flow energy equation and assuming $z_1 = z_2$, equation (2) gives

$$\text{Work done/unit mass} = w = (h_1 - h_2) + \tfrac{1}{2}(v_1{}^2 - v_2{}^2) + q \tag{3}$$

Since the process is adiabatic $q = 0$.

From tables:

$$\text{At } p_1 = 717\,\text{kN/m}^2, \quad t_1 = 866°\text{C}, \quad h_1 = 994\!\cdot\!4\,\text{kJ/kg}$$
$$p_2 = 217\,\text{kN/m}^2, \quad t_2 = 624°\text{C}, \quad h_2 = 716\!\cdot\!4\,\text{kJ/kg}$$
$$\text{Also } v_1 = 162\,\text{m/s}, \quad v_2 = 300\,\text{m/s}.$$

Substituting in equation (3),
Work done/unit mass $= w$

$$= 10^3 \times (994\!\cdot\!4 - 716\!\cdot\!4) + \tfrac{1}{2}(162^2 - 300^2)\,\text{J/kg}$$
$$= 278 \times 10^3 - 32 \times 10^3\,\text{J/kg}$$
$$= \mathbf{246\,kJ/kg}$$

6.3 Pitot-static tube

(a) Using the steady flow energy equation, in conjunction with the relation

$$\left(\frac{p_2}{p_1}\right) = \left(\frac{T_2}{T_1}\right)^{\gamma/(\gamma-1)}$$

which applies to isentropic changes in a perfect gas, show that the ratio of the stagnation pressure to static pressure is related to the Mach number M of a gas flow by the equation

$$\frac{p_0}{p} = \left[1 + \left(\frac{\gamma-1}{2}\right)M^2\right]^{\gamma/(\gamma-1)}$$

(b) A pitot-static tube inserted in a flow of argon gives a total pressure reading of 158 kN/m² abs and a static pressure of 104 kN/m² abs. The temperature of the gas is 20°C. Determine the speed of the gas flow, and the error which would occur in this determination if the gas were assumed to be incompressible with a density equal to that in the undisturbed stream. Instrumental errors are to be ignored. For argon, gas constant $R = 208 \cdot 2$ J/kg-K, ratio of specific heats $\gamma = 1 \cdot 68$.

Solution. (a) For the case in which no external work is involved the steady flow energy equation becomes

$$h + \tfrac{1}{2}v^2 + gz = h_0 + \tfrac{1}{2}v_0^2 + gz_0$$

At stagnation point $v_0 = 0$; also $z = z_0$. Therefore

$$h + \tfrac{1}{2}v^2 = h_0 \qquad (1)$$

For a perfect gas $h = C_p T$; also $R = C_p - C_v$

or $\qquad \dfrac{R}{C_p} = 1 - \dfrac{1}{\gamma} \qquad$ therefore $\qquad C_p = \dfrac{R\gamma}{\gamma-1}$

Thus $\qquad h = \dfrac{\gamma RT}{\gamma-1} \qquad$ and $\qquad h_0 = \dfrac{\gamma RT_0}{\gamma-1}$

Substituting in equation (1),

$$\frac{\gamma RT}{\gamma-1} + \tfrac{1}{2}v^2 = \frac{\gamma RT_0}{\gamma-1}$$

$$\frac{T_0}{T} = 1 + \tfrac{1}{2}v^2 \frac{(\gamma-1)}{\gamma RT} \qquad (2)$$

Now the velocity of sound $a = \sqrt{(\gamma RT)}$ and $\dfrac{T_0}{T} = \left(\dfrac{p_0}{p}\right)^{(\gamma-1)/\gamma}$

Substituting in equation (2),

$$\frac{p_0}{p} = \left(1 + \tfrac{1}{2}(\gamma-1)\frac{v^2}{a^2}\right)^{\gamma/(\gamma-1)}$$

Putting $v/a = M =$ Mach number,

$$\frac{p_0}{p} = [1 + \tfrac{1}{2}(\gamma - 1)M^2]^{\gamma/(\gamma-1)} \tag{3}$$

(b) From equation (3),

$$\left(\frac{p_0}{p}\right)^{(\gamma-1)/\gamma} = 1 + \tfrac{1}{2}(\gamma - 1)M^2$$

$$M^2 = \left\{\left(\frac{p_0}{p}\right)^{(\gamma-1)/\gamma} - 1\right\}\frac{2}{\gamma - 1}$$

$$M^2 = \left\{\left(\frac{158}{104}\right)^{0\cdot68/1\cdot68} - 1\right\}\frac{2}{0\cdot68}$$

$$= (1\cdot52^{0\cdot405} - 1) \times 2\cdot95$$

$$= (1\cdot1845 - 1) \times 2\cdot95 = 0\cdot544$$

$$M = \sqrt{0\cdot544} = 0\cdot737 = \frac{v}{a} = \frac{v}{\sqrt{(\gamma RT)}}$$

$$v = 0\cdot737\sqrt{(\gamma RT)} = 0\cdot737\sqrt{(1\cdot68 \times 208\cdot2 \times 293)}\,\text{m/s}$$

$$= 0\cdot737 \times 320 = \textbf{236}\,\textbf{m/s}$$

If the gas were assumed to be incompressible,

$$\tfrac{1}{2}\rho v^2 = p_0 - p$$

$$v^2 = \frac{2(p_0 - p)}{\rho} = \frac{2RT}{p}\,(p_0 - p)$$

$$v = \sqrt{\left\{2RT\left(\frac{p_0}{p} - 1\right)\right\}}$$

$$= \sqrt{\{2 \times 208\cdot2 \times 293(1\cdot52 - 1)\}} = 252\,\text{m/s}$$

$$\text{Error} = \frac{252 - 236}{236} = \frac{16}{236}$$

$$= \textbf{6}\cdot\textbf{8 per cent}$$

6.4   Venturi meter

Show how the Bernoulli equation for the flow of compressible gas

$$\frac{\gamma}{\gamma - 1}\frac{p}{\rho} + \tfrac{1}{2}v^2 = \text{constant}$$

is obtained, stating what assumptions have been made.

A venturi meter 400 mm diam at inlet and 125 mm diam at the throat is used for measuring the flow of air. The pressure and temperature at inlet were 138 kN/m² abs and 17°C when the pressure in the throat was 117 kN/m² abs. Calculate the flow in kg/s. Assume adiabatic expansion with $\gamma = 1\cdot4$ and that for air $R = 287$ J/kg-K. The coefficient of discharge for the meter was $0\cdot96$.

Solution. From example **6.1**, ignoring gravity forces, equation (4) gives

$$\frac{\gamma}{\gamma - 1}\frac{p}{\rho} + \tfrac{1}{2}v^2 = \text{constant}$$

Flow through a venturi meter can be considered as adiabatic. In Fig. 6.2 at the full bore section (1) the area is a_1, velocity v_1, pressure p_1 and the density ρ_1, the corresponding values at the throat section (2) being a_2, v_2, p_2 and ρ_2.

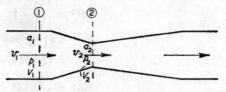

Figure   6.2

By Bernoulli's equation for adiabatic flow, example **6.1**(*b*),

$$\left(\frac{\gamma}{\gamma - 1}\right)\left(\frac{p_1}{\rho_1} - \frac{p_2}{\rho_2}\right) = \tfrac{1}{2}(v_2^2 - v_1^2) \tag{1}$$

Also since for adiabatic flow, $\dfrac{p_1}{\rho_1^{\gamma}} = \dfrac{p_2}{\rho_2^{\gamma}}$

$$\rho_2 = \left(\frac{p_2}{p_1}\right)^{1/\gamma}\rho_1 \quad \text{and} \quad \frac{p_2}{\rho_2} = \frac{p_1}{\rho_1}\left(\frac{p_2}{p_1}\right)^{(\gamma-1)/\gamma}$$

or putting $p_2/p_1 = r$,

$$\frac{p_2}{\rho_2} = \frac{p_1 r^{(\gamma-1)/\gamma}}{\rho_1} \tag{2}$$

For continuity of flow by mass

$$\rho_1 a_1 v_1 = \rho_2 a_2 v_2$$

$$v_2 = \frac{a_1}{a_2}\left(\frac{\rho_1}{\rho_2}\right)v_1$$

but

$$\frac{\rho_1}{\rho_2} = \left(\frac{p_1}{p_2}\right)^{1/\gamma} = \left(\frac{1}{r}\right)^{1/\gamma}$$

thus

$$v_2 = \frac{a_1}{a_2}\left(\frac{1}{r}\right)^{1/\gamma}v_1 \tag{3}$$

Substituting from equations (2) and (3) in equation (1)

$$\left(\frac{\gamma}{\gamma - 1}\right)\frac{p_1}{\rho_1}(1 - r^{(\gamma-1)/\gamma}) = \frac{v_1^2}{2}\left\{\left(\frac{a_1}{a_2}\right)^2\left(\frac{1}{r}\right)^{2/\gamma} - 1\right\}$$

$$v_1 = \sqrt{\frac{2\left(\dfrac{\gamma}{\gamma - 1}\right)\dfrac{p_1}{\rho_1}(1 - r^{(\gamma-1)/\gamma})}{\left(\dfrac{a_1}{a_2}\right)^2\left(\dfrac{1}{r}\right)^{2/\gamma} - 1}}$$

$$\text{Mass discharged per sec} = \dot{m} = C_d a_1 v_1 \rho_1$$

$$\dot{m} = C_d a_1 \rho_1 \sqrt{\frac{2\left(\dfrac{\gamma}{\gamma-1}\right)\dfrac{p_1}{\rho_1}(1 - r^{(\gamma-1)/\gamma})}{\left(\dfrac{a_1}{a_2}\right)^2 \left(\dfrac{1}{r}\right)^{2/\gamma} - 1}}$$

Note that this is only one form of the equation for $\dot{m}$ which could have been expressed in terms of p_2 and ρ_2 for example.

From the characteristic equation $p_1 = \rho_1 R T_1$ and so $\rho_1 = p_1/R T_1$

giving $\quad \dot{m} = \dfrac{C_d a_1 p_1}{R T_1} \sqrt{\dfrac{2\left(\dfrac{\gamma}{\gamma-1}\right) R T_1(1 - r^{(\gamma-1)/\gamma})}{\left(\dfrac{a_1}{a_2}\right)^2 \left(\dfrac{1}{r}\right)^{2/\gamma} - 1}}$

Putting $C_d = 0.96$, $\quad \gamma = 1.4$, $\quad R = 287\,\text{J/kg-K}$, $\quad T_1 = 17 + 273$
$$= 290\,\text{K}$$

$p_1 = 138 \times 10^3\,\text{N/m}^2\,\text{abs}$, $\quad p_2 = 117 \times 10^3\,\text{N/m}^2\,\text{abs}$

$$r = \frac{p_2}{p_1} = \frac{117}{138} = 0.85$$

$$d_1 = 400\,\text{mm}, \quad d_2 = 125\,\text{mm}, \quad \frac{a_1}{a_2} = \left(\frac{d_1}{d_2}\right)^2 = \left(\frac{400}{125}\right)^2 = 10.25$$

$$a_1 = \tfrac{1}{4}\pi d_1^2 = \tfrac{1}{4}\pi \times 0.16 = 0.126\,\text{m}^2$$

$$\dot{m} = \frac{0.96 \times 0.126 \times 138 \times 10^3}{287 \times 290} \times$$

$$\sqrt{\frac{2 \times \dfrac{1.4}{0.4} \times 287 \times 290(1 - 0.85^{0.4/1.4})}{(10.25)^2 \left(\dfrac{1}{0.85}\right)^{2/1.4} - 1}}$$

$$= 0.2006 \sqrt{\frac{582610(1 - 0.9546)}{105 \times 1.26 - 1}}\,\text{kg/s}$$

$$\dot{m} = 0.2006 \sqrt{\frac{26450}{131.3}} = \mathbf{2.85\,kg/s}$$

6.5 Orifice under maximum discharge conditions

Prove that when a compressible gas flows from a container under pressure, under conditions such that maximum flow is attained, the velocity of gas through the orifice is given by $v = \sqrt{(\gamma p/\rho)}$ where p and ρ are the pressure and density immediately in front of the orifice and the process follows the relation $p/\rho^\gamma = $ constant.

An air compressor which takes in $11.3\,\text{m}^3$ of air per minute at $101\,\text{kN/m}^2$ and $15°\text{C}$ is used to maintain $310\,\text{kN/m}^2$ gauge pressure in a large tank whence it flows back to atmosphere through an orifice with discharge coefficient 0.96. The temperature in the vessel is $23°\text{C}$. Calculate a suitable diameter for the orifice, $\gamma = 1.4$, and the gas constant $R = 287\,\text{J/kg-K}$.

Solution. Let p_1, v_1, ρ_1 be the pressure, velocity and density of the air inside the container (Fig. 6.3) and p_2, v_2, ρ_2 the corresponding values immediately outside the orifice. If the container is large $v_1 = 0$.

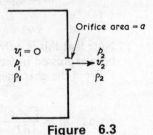

Figure 6.3

Flow through the orifice can be considered adiabatic and by Bernoulli's equation

$$\left(\frac{\gamma}{\gamma-1}\right)\left(\frac{p_1}{\rho_1}-\frac{p_2}{\rho_2}\right) = \frac{v_2^2 - v_1^2}{2} = \frac{v_2^2}{2}$$

For adiabatic flow
$$\frac{p_1}{\rho_1^{\gamma}} = \frac{p_2}{\rho_2^{\gamma}}$$

$$\rho_2 = \rho_1\left(\frac{p_2}{p_1}\right)^{1/\gamma} = \rho_1 r^{1/\gamma}$$

where $r = p_2/p_1$.

Hence
$$\left(\frac{\gamma}{\gamma-1}\right)\frac{p_1}{\rho_1}(1 - r^{(\gamma-1)/\gamma}) = \tfrac{1}{2}v_2^2 \qquad (1)$$

$$v_2 = \sqrt{2\left(\frac{\gamma}{\gamma-1}\right)\frac{p_1}{\rho_1}(1 - r^{(\gamma-1)/\gamma})}$$

Mass discharged/sec $= C_d a \rho_2 v_2 = C_d a \rho_1 r^{1/\gamma} v_2$

$$W = C_d a \rho_1 \sqrt{2\left(\frac{\gamma}{\gamma-1}\right)\frac{p_1}{\rho_1} r^{2/\gamma}(1 - r^{(\gamma-1)/\gamma})}$$

The mass discharged will clearly be a maximum for the value of r which makes $r^{2/\gamma}(1 - r^{(\gamma-1)/\gamma})$ a maximum, or when

$$\frac{d}{dr}\{r^{2/\gamma}(1 - r^{(\gamma-1)/\gamma})\} = 0$$

$$\frac{2}{\gamma}r^{(2-\gamma)/\gamma} - \frac{\gamma+1}{\gamma}r^{1/\gamma} = 0$$

$$r^{(\gamma-1)/\gamma} = \frac{2}{\gamma+1}$$

Under maximum flow conditions $\left(\frac{p_2}{p_1}\right)^{(\gamma-1)/\gamma} = \frac{2}{\gamma+1}$

Rewriting equation (1) in terms of the pressure p_2 and the density ρ_2 in the orifice

$$\tfrac{1}{2}v_2^2 = \left(\frac{\gamma}{\gamma-1}\right)\frac{p_2}{\rho_2}\left(\frac{1}{r^{(\gamma-1)/\gamma}} - 1\right)$$

For maximum flow $r^{(\gamma-1)/\gamma} = \frac{2}{\gamma+1}$

so that

$$\tfrac{1}{2}v_2{}^2 = \left(\frac{\gamma}{\gamma-1}\right)\frac{p_2}{\rho_2}\left(\frac{\gamma+1}{2}-1\right)$$

$$= \left(\frac{\gamma}{\gamma-1}\right)p_2 V_2 \left(\frac{\gamma-1}{2}\right) = \frac{\gamma p_2}{2\rho_2}$$

Velocity of gas in orifice $= v_2 = \sqrt{\dfrac{\gamma p_2}{\rho_2}}$

For air $\gamma = 1.4$ and the critical value of r is $0.528 = p_2/p_1$ and in this problem

$$p_1 = 310 + 101 = 411\,\text{kN/m}^2\ \text{abs}$$
$$p_2 = 101\,\text{kN/m}^2\ \text{abs}$$

so that $\qquad r = 101/411 = 0.246$

Since the pressure in falling from the inside value of $p_1 = 441\,\text{kN/m}^2$ to the outside value of $p_2 = 101\,\text{kN/m}^2$ must pass through a value p_t in the throat of the orifice for which $r = p_t/p_1$ would have the critical value, the orifice can be designed for maximum flow conditions with a velocity $v_t = \sqrt{(\gamma p_t/\rho_t)}$ in the throat,

or since $p_t/\rho_t = RT_t, \quad v_t = \sqrt{(\gamma RT_t)}$

where $T_t =$ throat temperature.

Now for adiabatic conditions $\dfrac{T_t}{T_1} = \left(\dfrac{p_t}{p_1}\right)^{(\gamma-1)/\gamma} = r^{(\gamma-1)/\gamma}$

Substituting $r^{(\gamma-1)/\gamma} = \dfrac{2}{\gamma+1}$ gives $\dfrac{T_t}{T_1} = \dfrac{2}{\gamma+1}$

Putting $T_1 = 23 + 273 = 296\,\text{K}$, $\gamma = 1.4$,

$$T_t = 296 \times \frac{2}{2.4} = 246.5\,\text{K}$$

and throat velocity $v_t = \sqrt{(\gamma RT_t)}$

$$= \sqrt{(1.4 \times 287 \times 246.5)}\,\text{m/s}$$
$$= 314.7\,\text{m/s}$$

If $d =$ orifice diameter and $Q =$ discharge

$$Q = C_d \tfrac{1}{4}\pi d^2 v_t$$

$$d = \sqrt{\frac{4Q}{C_d \pi v_t}}$$

Required $Q = \dfrac{11.3}{60}\,\text{m}^3/\text{s}$ at $101\,\text{kN/m}^2$ and $15°\text{C}$.

This must be corrected to throat temperature and pressure.

$$Q_t = \frac{11.3}{60} \times \frac{101}{p_t} \times \frac{T_t}{273+15}$$

$$p_t = 0.528 \times p_1 = 0.528 \times 411, \quad T_t = 246.5\,\text{K}$$

$$Q_t = \frac{11 \cdot 3}{60} \times \frac{101}{0 \cdot 528 \times 411} \times \frac{246 \cdot 5}{288} = 0 \cdot 075 \, \text{m}^3/\text{s}$$

Orifice diameter $d = \sqrt{\dfrac{4 \times 0 \cdot 075}{0 \cdot 96 \times \pi \times 314 \cdot 7}} = 0 \cdot 0178 \, \text{m} = \mathbf{17 \cdot 8 \, mm}$

6.6 Isothermal flow in pipes

The flow of gas through a pipe is isothermal. Assuming that the gain in kinetic energy of the gas is small in comparison with the energy absorbed in frictional resistance, show that the outlet pressure p_2 is given by

$$p_2 = p_1 \sqrt{\left(1 - \frac{4fLv_1^2}{DRT}\right)}$$

where p_1 is the inlet pressure, f the resistance coefficient, T the absolute temperature, and v_1 the inlet velocity. Hence calculate the outlet velocity when compressed air flows through at 100 mm diam pipe $1 \cdot 6$ km long if 7 m³/s enter the pipe at a pressure of 2000 kN/m². The temperature is constant at 15°C, $f = 0 \cdot 004$ and the gas constant $R = 287$ J/kg-K.

Solution. Since the pressure decreases along the pipe going towards the outlet the density of the gas will also decrease. The weight per second passing remains constant and the pipe area A is also constant.

For continuity of flow $\qquad \rho A v = \rho_1 A v_1$

or $\qquad\qquad\qquad\qquad \rho v = \rho_1 v_1$ \hfill (1)

so that as the density decreases the velocity increases. This has two effects:

(1) The kinetic energy of the gas must increase and part of the head available will be used in producing this increase.

(2) Since the velocity varies along the length of the pipe it is necessary to consider a small element of the length of the pipe when calculating the frictional resistance, integrating to obtain the total resistance.

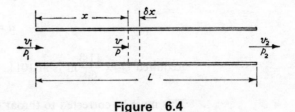

Figure 6.4

In Fig. 6.4 let the pressure be p at a distance x from the inlet and $p + \delta p$ at a distance $x + \delta x$. Using the Darcy formula and ignoring the change of kinetic energy due to change of density,

$$\delta p = -\rho g h_f = -\frac{\rho 4f\delta x v^2}{2D} \qquad (2)$$

From equation (1) $\qquad v = v_1 \dfrac{\rho_1}{\rho}$

where ρ is the mass density. For isothermal flow

$$\frac{p}{\rho} = \frac{p_1}{\rho_1}$$

so that $\qquad v = v_1 \dfrac{p_1}{p} \qquad (3)$

and $\qquad \rho = \dfrac{p}{p_1}\rho_1$

Substituting for v and ρ in equation (2)

$$\delta p = -\frac{p}{p_1}\rho_1 \frac{4f}{2D} v_1{}^2 \left(\frac{p_1}{p}\right)^2 \delta x$$

$$p\,dp = -\frac{2f\rho_1 p_1 v_1{}^2 dx}{D}$$

Integrating from $x = 0$, $\;p = p_1\;$ to $\;x = L$, $\;p = p_2$,

$$\frac{p_2{}^2 - p_1{}^2}{2} = -\frac{2f\rho_1 p_1 v_1{}^2 L}{D}$$

or since $\rho_1 = p_1/RT$

$$p_2{}^2 - p_1{}^2 = -\frac{4fp_1{}^2 v_1{}^2 L}{DRT}$$

$$p_2 = p_1 \sqrt{\left(1 - \frac{4fLv_1{}^2}{DRT}\right)}$$

Putting $\;p_1 = 2000\,\text{kN/m}^2$, $\;f = 0.004$, $\;L = 1600\,\text{m}$

$$D = 0.100\,\text{m}, \quad R = 287\,\text{J/kg-K}, \quad T = 288\,\text{K}$$

$$v_1 = \frac{Q}{\frac{1}{4}\pi D^2} = \frac{7}{\frac{1}{4}\pi \times 0.01 \times 60} = 14.9\,\text{m/s}$$

$$p_2 = 2000 \times 10^3 \sqrt{\left(1 - \frac{4 \times 0.004 \times 1600 \times (14.9)^2}{0.1 \times 287 \times 288}\right)}$$

$$= 2000 \times 10^3 \sqrt{(1 - 0.689)}\,\text{N/m}^2$$

$$= 2000 \times 10^3 \times 0.558 = 1116 \times 10^3\,\text{N/m}^2$$

$$= 1116\,\text{kN/m}^2$$

From equation (3) $v_2 = v_1 \dfrac{p_1}{p_2} = 14 \cdot 9 \times \dfrac{2000}{1116}$

$$= 26 \cdot 7 \, \text{m/s}$$

Problems

1 Prove that in the steady frictionless flow of a compressible fluid along a horizontal streamline

$$\int \frac{dP}{w} + \frac{v^2}{2g} = \text{constant}$$

where P, v and w denote pressure, velocity and specific weight of the fluid at any point.

Apply this equation to obtain the mass rate of flow of air in kg per min through a horizontal venturi meter having a 75 mm diam inlet and 25 mm diam throat, the absolute pressure at inlet and throat being respectively 125 kN/m² and 104 kN/m² and the air at the venturi inlet having a density of 1·5 kg/m³. Assume that flow is adiabatic taking $\gamma = 1 \cdot 41$.
Answer 6·7 kg/min

2 Calculate the mass of air flowing through a venturi meter having an inlet diameter of 100 mm and a throat diameter of 50 mm. The absolute pressures at inlet and throat were found to be 420 kN/m² and 350 kN/m² respectively. The temperature at inlet was 20°C. Assume that $R = 287$ J/kg-K and $\gamma = 1 \cdot 4$.
Answer 1·52 kg/s

3 Prove that in a frictionless adiabatic flow of a compressible fluid ($\gamma = 1 \cdot 4$) along a horizontal streamline

$$\left(\frac{\gamma - 1}{\gamma} \right) \frac{v^2}{2} + \frac{p}{\rho} = \text{constant}$$

where p, ρ, v denote respectively the pressure, mass density and velocity of the fluid.

An air stream in which the initial pressure is 101 kN/m² at 15°C and the velocity 105 m/s, is brought suddenly to rest by impinging on a solid obstacle. Calculate (*a*) the rise in pressure produced, (*b*) the change in temperature. For air $R = 287$ J/kg-K.
Answer 6·29 kN/m², 5·5°C

4 State Bernoulli's equation for the frictionless adiabatic flow of a gas and apply it to calculate the theoretical flow in kg/h of hydrogen gas through a horizontal venturi meter given the following information: diameter of meter at inlet = 75 mm and at throat = 25 mm; the pressure = 800 mm of mercury and the temperature = 15°C at inlet and the pressure = 765 mm of mercury at the throat. For hydrogen $PV^{1 \cdot 4} = $ constant for adiabatic expansion, and $R = 4110$ J/kg-K.
Answer 158 kg/h

5 Air flows through a venturi meter adiabatically. The diameter of the inlet is 75 mm and the pressure and temperature there are 138 kN/m² and 15°C. Find the diameter of the throat if the pressure is not to fall below 127·5 kN/m² when the discharge is 335 kg/h. Assume gas constant $R = 287$ J/kg-K, $\gamma = 1·4$ and $C_d = 1$.

Answer 19·25 mm

6 A horizontal pipe 200 mm diam conveys air at a pressure of 825 kN/m² abs and the maximum rate of flow is estimated to be 11 kg/s. If a venturi meter is to be fitted in the pipeline to measure the flow, find the least diameter at the throat which can be used if the pressure there is not to be less than 745 kN/m² abs. Assume that the temperature of the air at inlet to the meter is 20°C and for the constants R and γ use the values $R = 287$ J/kg-K and $\gamma = 1·4$. Work from first principles or prove any formula used.

Answer 107 mm

7 Flow of air along a pipe is measured by a venturi having an inlet diam of 75 mm and a throat diam of 50 mm. The temperature at entry to the meter is 15°C and the absolute pressure is 155 kN/m² while the absolute pressure at the throat is 112 kN/m². Assuming adiabatic conditions in the meter, and neglecting losses, find the rate of flow in kg/min. $R = 287$ J/kg-K.

Answer 42·2 kg/min

8 State Bernoulli's equation for the frictionless adiabatic flow of a gas. Thence deduce the equation for the theoretical mass rate of flow of a gas through a horizontal venturi meter having a ratio of the diameter at entry to the diameter at the throat represented by m.

Air is measured by a venturi meter having $m = 3$. The pressure difference between throat and inlet is one-tenth of the pressure at inlet. Find the percentage error in calculating the rate of flow on the assumption that density at the entrance is constant through the meter. Take $\gamma = 1·4$.

Answer 6 per cent.

9 A venturi meter having an inlet diameter of 75 mm and a throat diameter of 25 mm is used for measuring the rate of flow of air through a pipe. Mercury U-tube gauges register pressures at the inlet and the throat equivalent to 250 mm and 150 mm of mercury respectively. Determine the volume of air flowing through the pipe in dm³/s. Assume adiabatic conditions ($\gamma = 1·4$); density of air at inlet is 1·6 kg/m³. Barometric pressure is 760 mm of mercury.

Answer 96 dm³/s

10 Sketch a standard form of pitot-static tube and describe how such an instrument could be used to measure the rate of flow of dry air passing through a 1 m diameter trunk.

Calculate the local air velocity when the pitot-static reading was 6·5 mm water pressure. The pressure in the trunk was

103 kN/m² (abs) and the temperature 7·2°C. Dry air at 15°C and 100 kN/m² has a density of 1·21 kg/m³.

Answer 9·98 m/s

11 Air from a large vessel discharges into the atmosphere from a small orifice in its side. The pressure and temperature of the air in the vessel are 207 kN/m² abs and 15°C respectively. The diameter of the orifice is 25 mm. Assuming R and γ to be 287 J/kg-K and 1·4 respectively, calculate the mass of air discharging per s. The atmospheric pressure is 103·5 kN/m² and C_d for the orifice is 0·64.

Answer 0·154 kg/s

12 The quantity of air flowing in a pipeline is to be determined with the aid of an orifice. At a point upstream of the orifice the gas is under a pressure of 700 kN/m² abs and at a temperature of 38°C. What is the greatest downstream pressure for maximum velocity through the orifice? $\gamma = 1·404$.

Answer 369 kN/m²

13 Prove that the maximum continuous discharge of air through a convergent nozzle fitted in the side of a large vessel takes place when the pressure in the throat of the nozzle is 0·528 of the constant pressure of air in the vessel.

Find the diameter of a nozzle suitable for measuring the discharge from an air compressor which deals with 7 m³/min of atmospheric air at 101 kN/m² and 15°C. The nozzle is fitted into the side of a large air vessel into which the air is discharged from the compressor, and the pressure and temperature in the vessel are 227 kN/m² and 27°C. Assume a coefficient of discharge for the convergent nozzle of 0·99, R for air is 287 J/kg-K and $\gamma = 1·4$.

Answer 0·0187 m

14 A convergent-divergent nozzle is fitted into the side of a large vessel containing a gas under constant pressure and temperature. If the ratio of the specific heats of the gas is 1·3 to 1, calculate, from first principles, the percentage change in (*a*) pressure, (*b*) absolute temperature between the reservoir and the throat of the nozzle under maximum flow conditions. Neglect friction and assume adiabatic expansion.

Answer (*a*) 45·6 per cent, (*b*) 12 per cent

15 Show that Bernoulli's equation for compressible flow can be expressed as

$$\int \frac{dp}{w} + \frac{v^2}{2g} = \text{constant}$$

where the terms have the usual meaning.

A sharp-edged orifice 50 mm diam is used to measure the air intake of an engine. Obtain an expression for the volume of air per s passing through the orifice in terms of the pressure drop across the orifice measured in cm of water and the density of air in kg/m³ assuming it to be constant. The coefficient of discharge

for the orifice is $0 \cdot 61$. What would be the volume of air per minute corrected to 760 mm of Hg and 0°C if the pressure drop is 32 mm of water and the atmospheric pressure is 736 mm of mercury. Room temperature is 16°C; take $R = 287$ J/kg-K.
Answer $1 \cdot 514$ m³/min

16 Calculate the mass of air discharged per second into the atmosphere from a large vessel through a small orifice in its side if the pressure of air in the vessel is 172 kN/m² abs and its temperature 16°C. The diameter of the orifice is 25 mm and its coefficient of discharge is $0 \cdot 65$. The atmospheric pressure is 101 kN/m² abs and for air $R = 287$ J/kg-K. Assume adiabatic flow for which $\gamma = 1 \cdot 4$. Prove any formula you may use and work from Bernoulli's equation for compressible flow.
Answer $0 \cdot 01292$ kg/s

17 A nozzle is to be constructed to expand $4 \cdot 5$ kg/min of air from an initial pressure of 500 kN/m² abs to a final pressure of 150 kN/m² abs. Assuming that the initial temperature is $65 \cdot 6°C$ and the initial velocity negligible, find the temperature, velocity and cross-sectional area of the nozzle at intervals of 50 kN/m² pressure drop. Take $\gamma = 1 \cdot 4$.
Answer

$$T_2 = 338 \cdot 6 \left(\frac{p_2}{p_1} \right)^{0 \cdot 291}, v = \sqrt{677 \times 10^3 \left\{ 1 - \left(\frac{p_2}{p_1} \right)^{0 \cdot 291} \right\}},$$

$$a = \frac{21 \cdot 5 T_2}{p_2 v_2}$$

18 A convergent-divergent nozzle is fitted in the side of a tank containing air at constant pressure pN/m² and density ρkg/m³. The throat area of the nozzle is Sm².

Assuming that flow is adiabatic $(P = k\rho^\gamma)$ and frictionless, obtain an expression giving the mass rate of flow, i.e. kg/s in terms of p, ρ, S, γ and the pressure p_t at the throat of the nozzle.

Derive expressions, in terms of γ, giving the ratio between (*a*) the pressures and (*b*) the temperatures in the tank and at the throat when conditions are those corresponding to maximum discharge from the nozzle.

Answer (*a*) $\dfrac{p_t}{p} = \left(\dfrac{2}{\gamma + 1} \right)^{\gamma/(\gamma - 1)}$, (*b*) $\dfrac{T_t}{T} = \dfrac{2}{\gamma + 1}$

19 3m³/min of air at 1000 kN/m² is supplied to a pipeline 1800 m long and 75 mm diam. Calculate the pressure and volume delivered at the far end, assuming that the temperature, 17°C, is constant throughout and that only friction losses need be considered. Take $f = 0 \cdot 004$ and for air $R = 287$ J/kg-K.
Answer 630×10^3N/m², $4 \cdot 76$ m³/min

20 Hydrogen is being pumped along a pipe 150 mm diam and 300 mm long at a constant temperature of 25°C. The absolute pressure at exit and entry are respectively 102 kN/m² and 136 kN/m². Calculate the velocity at exit and the discharge in kg/min. Work from first principles and neglect the pressure drop due to the rate of change of momentum. Take the pipe friction coefficient as 0·004, and R for hydrogen as 4110 J/kg-K.

Briefly expain why the pressure drop due to the velocity change may be neglected without serious error.

Answer 172 m/s, 15·24 kg/min

21 A horizontal pipeline 900 m long and 75 mm diam conveys compressed air at a mean temperature of 18°C which may be assumed constant over the length of the pipe. If the gauge pressures at inlet and outlet of the pipe are 1240 kN/m² and 380 kN/m² respectively and the friction coefficient, assumed constant, is 0·0045, determine the flow through the pipe in kg/s. Any formula used must be proved. $R = 287$ J/kg-K for air and atmospheric pressure is 101 kN/m².

Answer 1·22 kg/s

22 Compressed air at 1100 kN/m² is supplied to a 100 mm diam main, 3·2 km in length. The temperature is constant throughout at 15°C and the volume supplied is 3·4 m³/min. Assuming $R = 287$ J/kg-K, and taking the coefficient of friction as 0·0040, calculate from first principles the pressure and velocity at the delivery end.

Answer 904 kN/m², 8·8 m/s

23 Calculate the terminal pressure and velocity of compressed air flowing through a horizontal 150 mm diam main 3·2 km long, if 8·4 m³/min enter the pipe at a pressure of 1700 kN/m². Assume that the temperature of the air is constant and equal to 15°C, $R = 287$ J/kg-K and $f = 0·0044$.

Answer 1435 kN/m², 9·42 m/s

7

Viscous flow

Fluid flow may be either viscous (laminar or streamline) or turbulent, the type of flow depending on the value of the Reynolds number. For flow in pipes the Reynolds number $Re = \rho v d / \mu$ where ρ = mass density and μ = dynamic viscosity of the fluid, v = mean velocity and d = pipe diameter. When Re is less than the critical value 2100 flow will be viscous; about 2100 there is a transitional region and for higher values of Re flow will always be turbulent.

For a given fluid in a pipe of given diameter there will therefore be a *critical velocity v*, corresponding to the critical Reynolds number 2100 below which flow is always viscous.

Since in viscous flow the frictional shear stress τ is equal to the dynamic viscosity μ multiplied by the velocity gradient (*see* Volume 1) problems of flow between parallel surfaces and in pipes can be solved theoretically.

Flow between parallel surfaces

7.1 Flow between parallel planes

> Stating carefully the main assumptions involved, develop, from first principles, an expression for the rate of steady flow, under laminar conditions, of a viscous incompressible fluid through a rectangular passage of width b, a very small depth h and of length L in the direction of flow, under a differential pressure p. Show also that the intensity of shear stress on the wall of the passage is $6\mu V / h$, where μ is the coefficient of viscosity and V is the mean velocity of flow. Neglect end and side effects.

Solution. Let the velocity of flow be v at a distance y from the centre-line of the passage (Fig. 7.1(*a*)) and consider the forces acting on an element of thickness $2y$ symmetrically disposed about the centre-line.

Force due to pressure difference causing motion

= pressure difference × cross-sectional area

$= (p_1 - p_2)b \times 2y = 2pby$

Neglecting the sides of the passage,

Force opposing motion due to viscosity

= viscous shear stress × area of upper and lower surfaces

$= \tau \times 2bL$

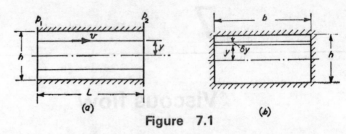

Figure 7.1

Now $\tau = \mu \times$ velocity gradient, and since the fluid in contact with the wall of the passage adheres to it and is at rest, v decreases as y increases and the velocity gradient is $- dv/dy$ so that

$$\text{Force opposing motion} = -\mu \frac{dv}{dy} \times 2bL$$

For steady flow without acceleration,

$$\text{Force causing motion} = \text{force opposing motion}$$

$$2pby = -\mu \frac{dv}{dy} \times 2bL$$

$$dv = -\frac{pydy}{\mu L} \tag{1}$$

Integrating

$$v = -\frac{py^2}{2\mu L} + A$$

To find the constant of integration A, we know that at the wall $v = 0$ when $y = \frac{1}{2}h$, hence

$$A = \frac{p}{2\mu L} \frac{h^2}{4}$$

and

$$v = \frac{p}{2\mu L}\left(\frac{h^2}{4} - y^2\right) \tag{2}$$

Note that the velocity distribution is a parabola; maximum velocity occurs at the centre-line when $y = 0$ and $v_{max} = ph^2/8\mu L$.

To find the discharge consider flow through an element of thickness δy (Fig. 7.1(b)).

$$\text{Discharge through element} = \delta Q = \text{area} \times \text{velocity}$$

$$\delta Q = b\delta y \times \frac{p}{2\mu L}\left(\frac{h^2}{4} - y^2\right)$$

on substituting for v from equation (2).

Integrating,

$$\text{Total discharge through passage} = Q = \frac{pb}{2\mu L} \int_{-h/2}^{+h/2} \left(\frac{h^2}{4} - y^2\right) dy$$

$$Q = \frac{pbh^3}{12\mu L}$$

To find the mean velocity V,

$$V = \frac{\text{discharge}}{\text{area}} = \frac{Q}{bh} = \frac{ph^2}{12\mu L} \tag{3}$$

To find the shear stress at the wall,

$$\text{Viscous shear } \tau = -\mu \frac{dv}{dy}$$

and from equation (1)

$$\frac{dv}{dy} = -\frac{py}{\mu L}$$

At the wall $\quad y = \frac{h}{2}$,

$$\frac{dv}{dy} = -\frac{ph}{2\mu L}$$

$$\text{Viscous shear stress at wall} = \frac{ph}{2L}$$

From equation (3)

$$p = \frac{12\mu L V}{h^2}$$

$$\text{Viscous shear stress at wall} = \frac{12\mu L V}{h^2} \times \frac{h}{2L} = \frac{6\mu V}{h}$$

7.2 Flow between parallel moving surfaces

Laminar flow of an incompressible fluid having viscosity μ takes place between two horizontal parallel surfaces whose distance apart is h, the upper surface being stationary while the lower surface moves with uniform velocity U. The length of the stationary surface in the direction of this velocity is L and a pressure drop P occurs over this length.

Assuming the surfaces have infinite width transverse to the direction of flow so that no end effects have to be considered, obtain the expression giving the quantity of fluid Q which passes under a transverse width B of the stationary surface.

Also obtain in terms of Q, B, h and μ the intensity of shear stress on the fluid adjacent to each surface.

Solution. Let τ be the viscous shear stress at a distance y from the lower surface at which the velocity is v and $(\tau + \delta\tau)$ the shear stress at $y + \delta y$ from the lower surface at which the velocity is $v - \delta v$ (Fig. 7.2). Considering the element shown of width B and thickness δy,

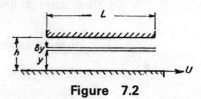

Figure   7.2

Viscous drag on element = surface area × difference of shear stress on upper and lower surface

$$= BL\delta\tau = BL\delta y \frac{d\tau}{dy}$$

Viscous shear stress $\tau = -\mu \dfrac{dv}{dy}$ since velocity gradient is negative

Thus
$$\frac{d\tau}{dy} = -\mu \frac{d^2v}{dy^2}$$

Viscous drag on element $= -BL\delta y\mu \frac{d^2v}{dy^2}$

Due to pressure difference

Force in direction of motion $= P \times B\delta y$

and for steady flow

Force in direction of motion = viscous drag

$$PB\delta y = -BL\delta y\mu \frac{d^2v}{dy^2}$$

$$\mu \frac{d^2v}{dy^2} = -\frac{P}{L}$$

$$\mu \frac{dv}{dy} = -y\frac{P}{L} + C_1$$

$$\mu v = -\tfrac{1}{2}y^2\frac{P}{L} + C_1 y + C_2$$

Since $v = U$ when $y = 0$, $\quad C_2 = \mu U$

$v = 0$ when $y = h$, $\quad C_1 = -\left(\frac{\mu U}{h} - \frac{h}{2}\frac{P}{L}\right)$

Thus
$$v = -\frac{P}{2\mu L}(y^2 - hy) + \frac{U}{h}(h - y)$$

Rate of flow $Q = \int_0^h Bv\delta y$

$$= B\int_0^h \left\{-\frac{P}{2\mu L}(y^2 - hy) + \frac{U}{h}(h - y)\right\}dy$$

$$Q = B\left(\frac{Uh}{2} + \frac{h^3}{12\mu}\frac{P}{L}\right)$$

Shear stress on fixed surface $\tau_h = -\mu\left(\frac{dv}{dy}\right)_{y=h} = \frac{\mu U}{h} + \frac{h}{2}\frac{P}{L}$

Shear stress on moving surface $\tau_0 = -\mu\left(\frac{dv}{dy}\right)_{y=0} = \frac{\mu U}{h} - \frac{h}{2}\frac{P}{L}$

7.3 Dashpot

A dashpot consists of a cylinder 7 cm diam in which slides a piston 8 cm long having a radial clearance of 1 mm. The cylinder is filled with an oil of viscosity 1 poise. Calculate the velocity of the piston when acted upon by a load having a mass of 18 kg.

Solution. As the piston moves oil is forced to flow between the cylinder wall and the piston. Since the radial clearance is very small compared to the piston diameter the flow can be treated as if the surfaces were parallel.

From equation (3) of example **7.1**

$$V = \frac{ph^2}{12\mu L}$$

where V = mean velocity of flow of oil between piston and wall

p = pressure difference between two ends of piston

$$= \frac{\text{load}}{\text{piston area}} = \frac{18 \times 9\cdot81}{\frac{1}{4}\pi \times (0\cdot07)^2} = 45\,906\,\text{N/m}^2$$

h = radial clearance = $0\cdot1\,\text{cm} = 0\cdot001\,\text{m}$

$\mu = 1$ poise = $0\cdot1\,\text{kg/m-s}$, L = length of piston = 8 cm

$$= 0\cdot08\,\text{m}$$

$$V = \frac{45\,906 \times (0\cdot001)^2}{12 \times 0\cdot1 \times 0\cdot08} = 0\cdot478\,\text{m/s}$$

Volume of oil leaving cylinder/s = $V \times$ clearance area

$$= 0\cdot478 \times \pi \times 0\cdot07 \times 0\cdot001\,\text{m}^3/\text{s}$$

$$= 0\cdot000\,105\,\text{m}^3/\text{s}$$

$$\text{Piston velocity} = \frac{\text{volume leaving cylinder/s}}{\text{piston area}}$$

$$= \frac{0\cdot000\,105}{\frac{1}{4}\pi \times (0\cdot07)^2}$$

$$= 0\cdot0273\,\text{m/s} = \mathbf{2\cdot73\,cm/s}$$

7.4   Critical velocity

In connexion with fluid flow define (1) the critical velocity and (2) Reynolds number. If the critical velocity of water in a 50 mm diam pipe is $0\cdot049$ m/s, find the critical velocity of air in m/s in a pipe of 150 mm diam given that the density of water is 100 kg/m³ and of air $1\cdot2$ kg/m³; the coefficient of viscosity of water is $0\cdot012$ poises and of air $0\cdot00018$ poises.

Solution. Fluid flow may be either viscous or turbulent depending on the velocity and physical conditions. For a given fluid flowing under given conditions (e.g. in a pipe of known diameter) there will be a critical velocity below which the flow will always be viscous and above which the flow will normally be turbulent.

Reynolds found that the change from viscous to turbulent flow depended also on the viscosity μ and mass density ρ of the fluid and, in the case of pipe flow, on the pipe diameter d. If v = velocity of fluid

$$\frac{\rho v d}{\mu} = \text{Reynolds number}$$

If Reynolds number, for pipe flow, is less than about 2100, flow will always be viscous.

For the 50 mm pipe with water flowing at critical Reynolds number

$$v_1 = 0 \cdot 049 \, \text{m/s}, \quad \rho_1 = 1000 \, \text{kg/m}^3$$
$$\mu_1 = 0 \cdot 012 \, \text{poises} = 0 \cdot 0012 \, \text{kg/m-s}$$
$$d_1 = 50 \, \text{mm} = 0 \cdot 05 \, \text{m}$$

$$\text{Critical Reynolds number} = \frac{\rho_1 v_1 d_1}{\mu_1} = \frac{1000 \times 0 \cdot 049 \times 0 \cdot 050}{0 \cdot 0012}$$
$$= 2042$$

For the 15 cm diam pipe with air flowing the critical Reynolds number will also be 2042

$$\rho_2 = 1 \cdot 2 \, \text{kg/m}^3$$
$$\mu_2 = 0 \cdot 00018 \, \text{poises} = 0 \cdot 000018 \, \text{kg/m-s}, \quad d_2 = 15 \, \text{cm} = 0 \cdot 15 \, \text{m}$$

At the critical velocity $\dfrac{\rho_2 v_2 d_2}{\mu_2} = 2042$

$$v_2 = \frac{2042 \times 0 \cdot 000018}{1 \cdot 2 \times 0 \cdot 15} = 0 \cdot 2042 \, \text{m/s}$$

7.5 Viscous flow in pipes

Show from first principles that the loss of pressure due to laminar flow in a horizontal circular pipe is given by $p = 32\mu V L/d^2$ where μ is the coefficient of viscosity of the fluid, V the mean velocity of flow, L and d length and diameter of the pipe, and hence that under these conditions the coefficient f in the expression $fLV^2/2gm$ is

16/Reynolds number

Oil with viscosity $1 \cdot 44 \, \text{kg/m-s}$ or N-s/m^2 and specific gravity $0 \cdot 9$ flows in a pipe 25 mm diam and 3 m long at 1/500 of the critical speed for which the Reynolds number is 2500. Calculate the head in metres of oil required to maintain the flow.

Solution. Consider a cylindrical element of radius r in the fluid (Fig. 7.3(a)) and let the velocity be v at this radius. The forces acting are

Force causing motion due to pressure difference

$$= (p_1 - p_2) \times \pi r^2 = p\pi r^2$$

Viscous drag on cylindrical surface

$$= \text{area} \times \text{viscous shear stress}$$
$$= 2\pi r L \mu \, \frac{dv}{dr}$$

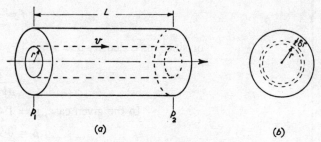

Figure 7.3

Since flow is steady there is no resultant force and

$$p\pi r^2 + 2\pi r L \mu \frac{dv}{dr} = 0$$

$$dv = -\frac{prdr}{2\mu L}$$

Integrating

$$v = -\frac{pr^2}{4\mu L} + A$$

At the wall $v = 0$ and $r = d/2$, thus $A = \frac{pd^2}{16\mu L}$

$$v = \frac{p}{4\mu L}\left(\frac{d^2}{4} - r^2\right)$$

Consider an annular element (Fig. 7.3(b)).

Flow through element $= \delta Q = $ area $\times$ velocity

$$\delta Q = 2\pi r \delta r \times v = \frac{2\pi p}{4\mu L}\left(\frac{d^2}{4} - r^2\right)r\delta r$$

Integrating, Total discharge $Q = \frac{2\pi p}{4\mu L}\int_0^{d/2}\left(\frac{d^2}{4} - r^2\right)r\,dr$

$$Q = \frac{\pi p d^4}{128\mu L}$$

Mean velocity $= V = \dfrac{Q}{\pi/4d^2} = \dfrac{pd^2}{32\mu L}$

Thus

$$p = \frac{32\mu L V}{d^2} \tag{1}$$

The expression $fLV^2/2gm$ is the Darcy formula for loss of head h_f, the term m being the hydraulic radius which is $d/4$ for a pipe, thus

$$\frac{fLV^2}{2gm} = \frac{4fLV^2}{2gd} = h_f = \frac{p}{w}$$

Substituting for p from equation (1),

$$\frac{4fLV^2}{2gd} = \frac{32\eta L V}{wd^2}$$

$$f = \frac{16 \mu g}{w V d}$$

or since mass density $\rho = w/g$

$$f = \frac{16}{\rho V d/\mu} = \frac{16}{\text{Reynolds number}}$$

In the given case, $\mu = 1{\cdot}44\,\text{kg/m-s}$

$$\rho = 0{\cdot}9 \times 10^3\,\text{kg/m}^3, \quad d = 0{\cdot}025\,\text{m}$$

Critical speed v_c corresponding to $Re = 2500$ is given by

$$\frac{\rho v_c d}{\mu} = 2500$$

$$v_c = \frac{2500 \times 1{\cdot}44}{0{\cdot}9 \times 10^3 \times 0{\cdot}025} = 160\,\text{m/s}$$

Mean velocity in pipe $= V = \dfrac{v_c}{500} = 0{\cdot}32\,\text{m/s}$

Head required to maintain flow = head lost in friction $= \dfrac{4fL}{d}\dfrac{V^2}{2g}$

Reynolds No. $= R = \dfrac{1}{500} \times 2500 = 5$

$$f = \frac{16}{R} = \frac{16}{5} = 3{\cdot}2, \quad L = 3\,\text{m}, \quad d = 0{\cdot}025\,\text{m}$$

Head required to maintain flow $= \dfrac{4 \times 3{\cdot}2 \times 3 \times (0{\cdot}32)^2}{0{\cdot}025 \times 2 \times 9{\cdot}81}$

$$= 8{\cdot}02\,\text{m}$$

7.6 Power required for viscous flow in pipeline

Calculate the power required to pump 50 metric tons of oil per hour along a pipeline 100 mm diam and $1{\cdot}6$ km long if the oil has a density of 915 kg/m³ and has a kinematic viscosity of $0{\cdot}00186$ m²/s.

Solution.

$$\text{Discharge} = Q = \frac{50 \times 1000}{915 \times 3600} = 0{\cdot}0152\,\text{m}^3/\text{s}$$

$$\text{Mean velocity } v = \frac{Q}{(\pi/4)d^2} = \frac{0{\cdot}0152}{(\pi/4) \times (0{\cdot}1)^2} = 1{\cdot}94\,\text{m/s}$$

$$\text{Reynolds number} = \frac{vd}{\nu} = \frac{1{\cdot}94 \times 0{\cdot}1}{0{\cdot}00186} = 104$$

Therefore the flow is laminar and $f = \dfrac{16}{Re} = \dfrac{16}{104} = 0{\cdot}154$.

$$\text{Head lost in friction} = \frac{4fL}{d}\frac{v^2}{2g} = \frac{4 \times 0\cdot154 \times 1\cdot6 \times 10^3 \times (1\cdot94)^2}{0\cdot1 \times 2g}$$

$$= 1891\,\text{m of oil}$$

$$\text{Power required} = \text{wt/s} \times \text{head lost}$$

$$= \frac{50 \times 1000 \times 9\cdot81}{3600} \times 1891\,\text{watts}$$

$$= \mathbf{258kW}$$

7.7 Flow through annular space

Derive the equation

$$\frac{d}{dr}\left(r\,\frac{dv}{dr}\right) = \frac{1}{\mu}\,r\,\frac{dp}{dx}$$

for axial viscous flow along a circular pipe, where v is the viscosity at radius r, μ is the viscosity of the fluid, and dp/dx is the pressure gradient along the pipe.

Hence show, by integrating the above expression, that if viscous flow takes place in the annular space between two concentric pipes of radii R_1 and R_2 $(R_1 < R_2)$, the maximum velocity occurs at a radius

$$R = R_1 \sqrt{\frac{a^2 - 1}{2\log_e a}}$$

where $a = R_2/R_1$.

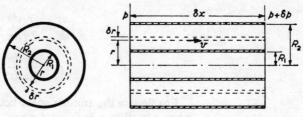

Figure 7.4

Solution. In Fig. 7.4 consider an annular element of radius r and thickness δr whose velocity is v and let the pressures at two sections δx apart be p and $p + \delta p$. Then if the pressure gradient is dp/dx,

$$\delta p = \delta x\,\frac{dp}{dx}$$

Force due to pressure difference = area of annulus $\times$ δp

$$= 2\pi r\delta r\,.\,\delta x\,\frac{dp}{dx}$$

This acts from right to left.

Viscous drag on element will be the difference of the shear forces on the inside and outside of the cylindrical shell of the element. At any radius r

Viscous shear force = surface area × viscous shear stress

$$= 2\pi r\delta x \times \tau = -2\pi r\delta x . \mu \frac{dv}{dr}$$

Differentiating

Rate of change of shear force with radius $= -2\pi\mu\delta x \frac{d}{dr}\left(r\frac{dv}{dr}\right)$

Multiplying by the thickness of the element δr

Viscous drag on element $= -2\pi\mu\delta x . \frac{d}{dr}\left(r\frac{dv}{dr}\right) . \delta r$

This also acts from right to left.

Since the fluid is not accelerating the sum of the pressure force and viscous drag is zero.

$$2\pi r\delta r\delta x \frac{dp}{dx} - 2\pi\mu\delta x \frac{d}{dr}\left(r\frac{dv}{dr}\right)\delta r = 0$$

Thus
$$\frac{d}{dr}\left(r\frac{dv}{dr}\right) = \frac{1}{\mu} r \frac{dp}{dx} \qquad (1)$$

Note. The sign convention in this problem differs from that in example 7.5 since it has been assumed that p increases with x.

The velocity at any radius r is obtained by integrating equation (1) twice.

$$r\frac{dv}{dr} = \frac{r^2}{2\mu}\frac{dp}{dx} + A$$

$$\frac{dv}{dr} = \frac{r}{2\mu}\frac{dp}{dx} + \frac{A}{r}$$

$$v = \frac{r^2}{4\mu}\frac{dp}{dx} + A\log_e r + B \qquad (2)$$

For flow in the annular space between two tubes of radius R_1 and R_2, when $r = R_1$, $v = 0$ and when $r = R_2$, $v = 0$.
Substituting in equation (2),

$$A\log_e R_1 + B = -\frac{R_1^2}{4\mu}\frac{dp}{dx}$$

$$A\log_e R_2 + B = -\frac{R_2^2}{4\mu}\frac{dp}{dx}$$

$$A(\log_e R_2 - \log_e R_1) = \frac{R_1^2 - R_2^2}{4\mu}\frac{dp}{dx}$$

$$A = \frac{R_1^2 - R_2^2}{4\mu}\frac{dp}{dx}\frac{1}{\log_e R_2/R_1}$$

The velocity v is a maximum when $\dfrac{dv}{dr} = 0$.

$$\frac{dv}{dr} = \frac{r}{2\mu}\frac{dp}{dx} + \frac{A}{r} = 0$$

$$\frac{r^2}{2\mu}\frac{dp}{dx} + \frac{R_1^2 - R_2^2}{4\mu}\frac{dp}{dx}\frac{1}{\log_e R_2/R_1} = 0$$

$$r^2 = \frac{R_1^2}{2}\left(\frac{R_2^2}{R_1^2} - 1\right)\frac{1}{\log_e R_2/R_1}$$

For maximum velocity

$$r = R_1\sqrt{\frac{a^2 - 1}{2\log_e a}} \quad \text{where} \quad a = \frac{R_2}{R_1}$$

Problems

1 Glycerine is forced through the narrow space formed between two plane glass plates. The distance between the plates is t and the width of the space is b. Assuming that flow is laminar, two-dimensional and fully developed, derive an expression for the longitudinal pressure gradient in terms of the volume rate of flow Q, the fluid viscosity μ, and the distances b and t.

In a particular version of the Hele-Shaw analogy the above conditions apply with $b = 0\cdot10$ m and $t = 0\cdot00076$ m. The fluid is glycerine with a coefficient of viscosity of $0\cdot96$ N-s/m². A pressure difference of 192 kN/m² is applied over a length of $0\cdot23$ m. Determine the maximum velocity occurring in the centre plane of the space.
Answer $0\cdot063$ m/s

2 The radial clearance between a plunger and the walls of a cylinder is $0\cdot075$ mm, the length of the plunger is 250 mm and its diameter 100 mm. There is a difference in pressure of the water on the two ends of the plunger of 207 kN/m² and the viscosity of the water is $1\cdot31 \times 10^{-3}$ kg/m-s. Treating the flow as if it occurred between parallel flat plates, estimate the rate of leakage in litres/s.
Answer $6\cdot98 \times 10^{-3}$ litres/s

3 A storage tank containing oil of viscosity $0\cdot7$ poises is cylindrical with its axis vertical and is 6 m in diameter. When the oil is under pressure at 345 kN/m², leakage occurs at a circumferential seam which consists of a riveted lap joint. The effective gap between the plates is found to be $0\cdot025$ mm, the plates overlap 100 mm and the rivets reduce the effective circumferential length of the opening by 40 per cent. Calculate the rate of leakage in dm³/h.
Answer $2\cdot61$ dm³/h

4 A dashpot consists of a piston $143\cdot5$ mm diam working concentrically in a cylinder of $143\cdot6$ mm bore. The cylinder contains oil of viscosity $0\cdot8$ poises. Calculate the force which must be

applied to the piston to give it a velocity of $0 \cdot 003$ m/s, when 250 mm of piston is in the cylinder.
Answer 3336 kN

5 The radial clearance between an hydraulic plunger and the cylinder wall is $0 \cdot 1$ mm, the length of the plunger $0 \cdot 3$ m and the diam 100 mm. Find the velocity and rate of leakage past the plunger at an instant when the difference of pressure between the two ends of the plunger is 9 m of water. Take $\mu = 1 \cdot 31 \times 10^{-3}$ kg/m-s.
Answer $0 \cdot 188$ m/s, $5 \cdot 9 \times 10^{-6}$ m³/s

6 Water at 20°C leaks through a horizontal slot $0 \cdot 25$ mm deep, 100 mm broad and 150 mm long. If the pressure difference is 34 kN/m² what will be the rate of leakage?
Answer $0 \cdot 105$ m³/h

7 Viscous flow takes place between two stationary parallel plates whose length in the direction of flow is L and distance apart is h.
Show that if the width of the plates is large compared with h so that end and side effects can be neglected, the pressure drop in the direction of flow is

$$p = \frac{12 \mu L v}{h^2}$$

where μ = viscosity of the fluid, v = mean velocity of the fluid.
A piston of diameter a and length L moves concentrically in an oil dashpot, the small radial clearance being h. If the piston is moving downward at a velocity v_p find the resistance to motion due to (a) the pressure on the underside of the piston, and (b) the viscous shear on the walls of the piston.
Find the ratio of radial clearance to piston diameter in order that the resistance due to (b) shall be 1 per cent of that due to (a). Assume that the piston velocity is inappreciable when compared with the mean velocity of the fluid.
Answer 1 to 200

8 Oil having a coefficient of dynamic viscosity of $0 \cdot 083$ kg/m-s flows between two very large parallel flat plates 24 mm apart. If the mean velocity of the oil is $0 \cdot 15$ m/s, what is the shearing stress at 6 mm and 12 mm from the lower plate?
Answer $1 \cdot 56$ N/m², 0

9 Oil of viscosity $0 \cdot 8$ poises leaks from a container through a joint which is $0 \cdot 6$ m wide, 50 mm long in the direction of flow, with a gap between the parallel surfaces of $0 \cdot 25$ mm. Calculate the volume of oil escaping per hour if the pressure difference between inside and outside is 35 kN/m².
Answer $24 \cdot 6$ dm³/h

10 Compare the friction losses in a 25 mm diam pipe when water, coefficient of viscosity $0 \cdot 013$ poises (c.g.s. units), flows at the rate of (1) 160 dm³/h, (2) 680 dm³/h given that the frictional

coefficient f is $16R^{-1}$ for streamline flow and $0 \cdot 064 R^{-0 \cdot 23}$ for turbulent flow, where R is the Reynolds number (vd/v).
Answer   1 to 16·2

11   Define the coefficients of viscosity and kinematic viscosity and give the dimensions of each.

Water of density 1000 kg/m³ and coefficient of viscosity $0 \cdot 012$ poises (c.g.s. units) flows along a pipe of 50 mm diam and the measured discharge is $2 \cdot 8$ dm³/s. Given that the frictional coefficient f is $0 \cdot 064R^{-0 \cdot 23}$ where R is the Reynolds number (vd/v) find the pressure drop in N/m² over a length of 6 m of pipe.
Answer   12·2 kN/m²

12   An oil of mean density 880 kg/m³ flows under a head of 30 m through 3000 m of pipe, $0 \cdot 3$ m in diam. Due to cooling the viscosity changes along the length and may be taken as $0 \cdot 57$ kg/m-s over the first 1500 m and $1 \cdot 14$ kg/m-s over the second 1500 m. Verify that laminar flow conditions exist and determine the flow in dm³/s, neglecting entry and exit losses.
Answer   20·1 dm³/s

13   Oil of viscosity $0 \cdot 048$ kg/m-s flows through an 18 mm diam pipe with a mean velocity of $0 \cdot 3$ m/s. Calculate the pressure drop which occurs over a length of 45 m of pipe.

Calculate also the velocity at a distance of 3 mm from the wall of the pipe.
Answer   64 kN/m², 0·332 m/s

14   An oil having an absolute viscosity of $0 \cdot 048$ kg/m-s flows through a 25 mm diam pipe at an average velocity of $0 \cdot 3$ m/s. Calculate the pressure drop in 30 m of pipe and the velocity at a distance of 6 mm from the wall of the pipe.
Answer   22·1 kN/m², 0·463 m/s

15   Oil having a specific gravity of $0 \cdot 85$ is pumped through a horizontal pipe 150 mm diam and 1200 m long. The quantity of oil discharged is 23 dm³/s when the pump, which has an efficiency of 65 per cent, requires $7 \cdot 5$ kW to drive it.

Taking the friction coefficient f in the expression $4flv^2/2gd$ as being equal to $16(N)^{-1}$ where N is the Reynolds number, obtain the viscosity and kinematic viscosity of the oil, expressing the former in poises and the latter in stokes.

Also explain why it is justifiable to use the friction coefficient in the form given.
Answer   0·955 poises, 1·123 stokes

16   A pipe 75 mm diam and 900 mm long conveys oil of sp. gr. $0 \cdot 85$ and kinematic viscosity $0 \cdot 0033$ m²/s at the rate of 40 metric tons/h. Confirm that the flow is laminar and find the power absorbed in overcoming friction in the pipe.
Answer   55·6 kW

17   The velocity along the centre-line of a 150 mm diam pipe conveying oil under laminar flow conditions is 3 m/s. The viscosity of the oil is $1 \cdot 2$ poises and its specific gravity is $0 \cdot 9$.

Assuming that the velocity distribution across the pipe is parabolic, obtain: (a) the quantity flowing in dm³/s, (b) the shear stress in the oil at the pipe wall in N/m². Also verify that the flow is laminar.

Answer 26·5 dm³/s, 9·6 N/m²

18 Explain what is meant by the term "critical velocity" as applied to the flow of a fluid in a pipe running full.

Oil of specific gravity 0·9 and viscosity 0·17 kg/m-s is pumped through a 75 mm diam pipe 750 m long at the rate of 2·75 kg/s. If the critical Reynolds number is 2000, show that the critical velocity is not exceeded and calculate the pressure required at the pump and the power required. Prove any formula used.

Answer 502 kN/m², 1·53 kW

19 A circular pipe of radius R_1 is placed concentrically inside another circular pipe, the larger pipe having an inside radius of R_2. Considering the case of axial laminar flow of an incompressible fluid in the annular space between the pipes, show that the maximum velocity occurs at a radius

$$R_0 = \sqrt{\frac{R_2^2 - R_1^2}{2 \log_e R_2/R_1}}$$

For the particular case of $R_2 = 2R_1$, find the ratio of R_0 to R_1 and express the velocity at a radius $1·25R_1$ as a ratio of the maximum velocity.

Answer 1·47, 0·796

20 An oil cooler consists of tubes of 12 mm internal diam 3·5 m long, and the oil of specific gravity 0·90 is forced through at a speed of 1·8 m/s. The coefficient of viscosity at the inlet is 0·28 c.g.s. units and at the outlet is 1·0 c.g.s. units; it may be taken to vary as a linear function of the length. Estimate the power required to force the oil through a group of 200 tubes. Establish any formula used.

Answer 3·65 kW

8

Turbulent flow

Much of the work of hydraulic engineers is concerned with the flow of liquids and gases under turbulent flow conditions for which in pipe flow the value of Reynolds number exceeds 2500. The particles of fluid do not move in a uniform orderly manner but have small irregular motions, superimposed on their main motion, which have velocity components at right angles to the direction of flow. Particles are thus continuously being interchanged between adjoining layers in the fluid, producing eddies, and the accompanying interchange of momentum will produce shear stresses between layers moving at different velocities.

The frictional resistance to flow in pipes can be estimated by using the Darcy formula $h_f = (4fL/d)(v^2/2g)$.

Investigation of turbulent flow in pipes has been largely concerned with the determination of the value of the friction coefficient f and its variation with Reynolds number and pipe roughness.

8.1 Resistance coefficients

> Fluid of mass density ρ flows through a pipe of diameter d with a mean velocity V. Establish from first principles a relationship between the shear stress τ_0 between the fluid and the wall of the pipe and the loss of head per unit length i. Obtain also a relationship between τ_0 and the resistance coefficient f in the Darcy formula.
>
> How does the shear stress in the fluid vary across the pipe?

Solution. The pressures at two points, distance L apart in a pipe of radius R, are p_1 and p_2 (Fig. 8.1). If the fluid flows with a steady mean velocity V then

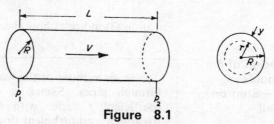

Figure 8.1

Accelerating force due to pressure difference
      = retarding force due to shear stress on pipe wall

$$(p_1 - p_2) \times \pi R^2 = 2\pi RL \times \tau_0$$

$$\tau_0 = \frac{p_1 - p_2}{L} \frac{R}{2} = \frac{\rho g h_f}{L} \frac{R}{2}$$

$$\tau_0 = \tfrac{1}{2}\rho g i R$$

or if m = hydraulic radius = $\tfrac{1}{2}R$

$$\tau_0 = \rho g i m \qquad (1)$$

The Darcy formula states

$$h_f = \frac{4fL}{D} \frac{V^2}{2g} = \frac{fL}{m} \frac{V^2}{2g}$$

$$\frac{h_f}{L} = i \quad \text{so that} \quad mig = \tfrac{1}{2}fV^2$$

Substituting in equation (1)

$$\tau_0 = \tfrac{1}{2}\rho f V^2$$

$$f = 2\frac{\tau_0}{\rho V^2} = 2\frac{mig}{V^2} \qquad (2)$$

At any other radius r suppose the shear stress in the fluid is τ. Considering the forces on a cylinder of fluid of radius r and length L, if there is no acceleration

Force due to pressure difference = force due to shear stress

$$(p_1 - p_2) \times \pi r^2 = 2\pi rL \times \tau$$

$$\tau = \frac{p_1 - p_2}{L} \frac{r}{2}$$

But as shown above

$$\tau_0 = \frac{p_1 - p_2}{L} \frac{R}{2}$$

Thus

$$\frac{\tau}{\tau_0} = \frac{r}{R}$$

If y = distance from wall = $R - r$

$$\tau = \tau_0 \left(\frac{R - y}{R} \right)$$

$$\tau = \tau_0 (1 - y/R) \qquad (3)$$

Equation (3) indicates how shear stress τ varies across the pipe.

8.2 Variation of f with Reynolds Number – Stanton and Pannell

Give an account of the work of Stanton and Pannell on flow through pipes. Sketch a curve showing how the friction coefficient f varies with Reynolds number, covering both streamline and turbulent flow.

Solution. Sir T. E. Stanton and J. R. Pannell carried out extensive tests with water, air and oil through pipes varying from capillary tubes

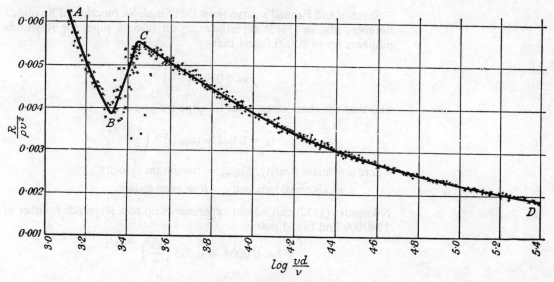

Figure 8.2

to water mains 18 ft (5·5 m) in diameter to confirm Reynolds' law of similarity.

The results of these experiments were published in 1914 and give a striking confirmation of the law of similarity for pipes for, as seen in Fig. 8.2, all the observations conformed very closely to a single curve when the resistance coefficient

$$\frac{R}{\rho v^2} \left(\text{which is } \frac{\tau_0}{\rho V^2} = \frac{f}{2} \right)$$

is plotted on a base of the logarithm of the pipe Reynolds number

$$\left(\log_{10} \frac{vd}{v} \right).$$

This curve shows clearly the two different types of flow. From A to B the curve shows the results for laminar or viscous flow, B is the point corresponding to the lower critical velocity and a Reynolds number of 2100. From B to C there is a transitional region, C corresponding to the higher critical velocity, and the curve from C to D represents turbulent flow conditions.

Darcy had originally proposed for turbulent flow that the value of f depended on the pipe diameter d and on the condition of the pipe, giving

$$f = 0.005 \left(1 + \frac{1}{12d} \right) \text{ for new smooth pipes}$$

$$f = 0.01 \left(1 + \frac{1}{12d} \right) \text{ for old worn pipes}$$

where d is measured in feet.

Stanton and Pannell's curve show that f is also a function of Reynolds number. Blasius (1913) experimenting on smooth pipes for Reynolds numbers up to 80000 found that

$$f = 0.0791 \left(\frac{Vd}{\nu}\right)^{-1/4}$$

and that the shear stress τ_0 at the pipe wall was

$$\tau_0 = 0.0225\rho \, (u_{\text{max}})^{7/4} \left(\frac{\nu}{R}\right)^{1/4}$$

where ρ = mass density, u_{max} = maximum velocity,
ν = kinematic viscosity, R = pipe radius.

Nikuradse (1932) carried out experiments up to a Reynolds number of 3240000 and found that

$$f = 0.0008 + 0.055 \left(\frac{Vd}{\nu}\right)^{-0.237}$$

Other relations given are

$$f = 0.064 \left(\frac{Vd}{\nu}\right)^{-0.23} \qquad \frac{mig}{V^2} = 0.032 \left(\frac{Vd}{\nu}\right)^{-0.23}$$

8.3 Dimensional analysis

Show by dimensional analysis that for turbulent flow in rough pipes the resistance coefficient f is a function of Reynolds number and the ratio R/k where R is the pipe radius and k the average height of the roughness projections on the wall of the pipe.

Solution. It might reasonably be expected that the shear stress at the pipe wall $\tau_0[ML^{-1}T^{-2}]$ will depend on the mean velocity $V[LT^{-1}]$, mass density $\rho[ML^{-3}]$, pipe radius $R[L]$, coefficient of dynamic viscosity $\mu[ML^{-1}T^{-1}]$ and the height of the roughness projections $k[L]$.

Assume a simple relationship $\tau_0 = AV^a\rho^bR^c\mu^ek^g$ in which a, b, c, e and g are unknown indices and A a constant. Equating dimensions

$$ML^{-1}T^{-2} = L^aT^{-a} \cdot M^bL^{-3b} \cdot L^c \cdot M^eL^{-e}T^{-e} \cdot L^g$$

Equating powers of M, L and T

$$1 = b + e$$
$$-1 = a - 3b + c - e + g$$
$$-2 = -a - e$$

Solve for a, b and c.

$$a = 2 - e, \quad b = 1 - e, \quad c = -e - g$$

Substituting in the original equation

$$\tau_0 = AV^{2-e}\rho^{1-e}R^{-e-g}\mu^e k^g$$

$$\tau_0 = AV^2\rho\left(\frac{\rho VR}{\mu}\right)^e\left(\frac{R}{k}\right)^{-g}$$

or since $R = \frac{1}{2}d$ where d is the pipe diameter

$$\tau_0 = V^2\rho\phi\left\{\frac{\rho Vd}{\mu}, \quad \frac{R}{k}\right\}$$

where ϕ is "a function of".

The resistance coefficient $f = 2\dfrac{\tau_0}{\rho V^2}$

Thus $\qquad\qquad\qquad\qquad f = \phi\left\{\dfrac{\rho Vd}{\mu}, \quad \dfrac{R}{k}\right\}$

where $\rho Vd/\mu$ is the pipe Reynolds number.

8 .4 Effect of roughness, Nikuradse's experiments

Describe the results of Nikuradse's experiments on the effect of the relative roughness R/k on the value of the resistance coefficient f for turbulent flow in circular pipes.

Solution. Nikuradse's experiments (1933) were made on pipes of 2·5 cm, 5 cm and 10 cm diameter artificially roughened by gluing sand of 0·1 to 1·6 mm diameter to the walls internally. The relative roughness of the surface was measured by the ratio R/k where R is the pipe radius and k the average height of the roughness projections. The resistance to flow was measured for a range of values of the mean velocity and the value of the resistance coefficient f in the Darcy formula determined. Figure 8.3 shows the curves obtained when $\log_{10} f$ is plotted against $\log_{10}\rho Vd/\mu$ the logarithm of the pipe Reynolds number.

The straight line AB represents the laminar flow region (slope -1, corresponding to $f = 16/Re$) and the line CD represents turbulent flow in smooth pipes following Blasius' equation ($f = 0{\cdot}0791\ Re^{-1/4}$). Depending on the relative roughness, the curves for each value of R/k break away from the smooth pipe law CD starting with the roughest pipe (lowest R/k value). Above certain Reynolds number for each value of R/k the value of f becomes constant and the pressure drop in the pipe is proportional to the square of the velocity.

The curve shows that laminar flow is not affected by normal pipe roughness. The explanation offered for the effect of roughness on turbulent flow is that a boundary layer exists at the wall of the pipe in which flow is laminar. As long as the roughness projections do not penetrate this layer they cannot affect the main flow in the pipe and the pipe is effectively smooth. The thickness of the boundary layer decreases as velocity, and therefore Reynolds number, increases and eventually the projections penetrate the boundary layer and the curves in Fig. 8.3 break away from the smooth pipe law.

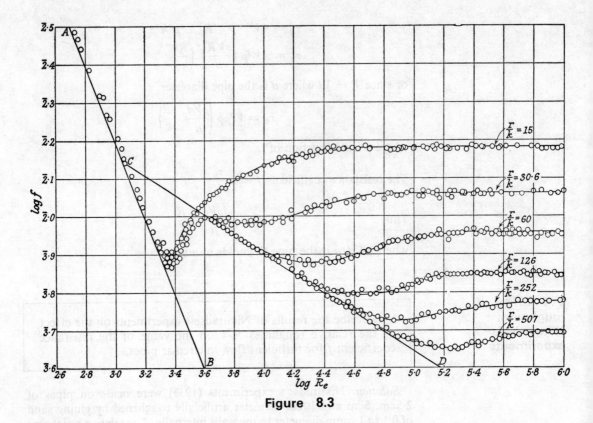

Figure 8.3

8.5 Velocity distribution, one-seventh power law

Sketch the normal velocity distribution across a pipe which may be expected for (a) laminar flow, (b) turbulent flow in smooth pipes, (c) turbulent flow in rough pipes.

Assuming that the shear stress at the wall is independent of pipe radius, that

$$f = 0 \cdot 0791 \left(\frac{Vd}{v} \right)^{-0 \cdot 25}$$

and that the ratio u/V is the same for all values of the mean velocity V where u is the velocity at a distance y from the pipe wall, show that the velocity distribution is given by

$$u = \text{constant} \times V \left(\frac{y}{R} \right)^{1/7}$$

where R is the radius of the pipe.

Solution. Figure 8.4 shows typical velocity distributions under normal conditions in a straight pipe for the three cases. The shape of the velocity distribution curves at a given section would be affected by any disturbance (obstruction, bend, etc.) immediately upstream. Such dis-

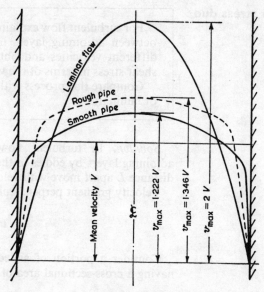

Figure 8.4

turbances gradually fade out along the pipe and the velocity distribution reverts to normal but the length required for this to occur may be as much as 30 to 40 pipe diameters.

Assuming that u/V is the same for all values of the mean velocity, it will be a function of y/R, or

$$u = CV\left(\frac{y}{R}\right)^n \quad \text{where } C \text{ and } n \text{ are constants}$$

Since

$$f = \frac{2\tau_0}{\rho V^2}$$

$$\tau_0 = \tfrac{1}{2}\rho V^2 f = \tfrac{1}{2}\rho V^2 \times 0.0791\left(\frac{Vd}{\nu}\right)^{-1/4}$$

$$= \text{constant} \times \rho V^2\left(\frac{\mu}{\rho V \times 2R}\right)^{1/4}$$

where μ = dynamic viscosity

$$\tau_0 = \text{constant} \times \rho^{3/4}V^{7/4}\mu^{1/4}R^{-1/4}$$

Since $u = CV\left(\dfrac{y}{R}\right)^n$, $V = \text{constant} \times u\left(\dfrac{y}{R}\right)^{-n}$

Thus $\quad \tau_0 = \text{constant} \times \rho^{3/4}u^{7/4}y^{-7n/4}\mu^{1/4}R^{(7n/4-1/4)}$

If the shear stress at the wall τ_0 is assumed independent of R, the index of R is zero, and

$$\frac{7n}{4} - \frac{1}{4} = 0 \quad \text{and} \quad n = \frac{1}{7}$$

Thus $\quad u = \text{constant} \times V\left(\dfrac{y}{R}\right)^{1/7}$

8.6 Shear stress due to eddies

For turbulent flow explain how momentum may be transferred between adjoining layers in the fluid which are moving with different velocities and obtain an expression for the resulting shear stress in terms of the velocity gradient.

Compare this process with that occurring when flow is viscous or laminar.

Solution. In turbulent flow momentum is transferred between adjoining layers by eddies in the fluid. In Fig. 8.5 two layers of fluid a distance L apart move with velocities u and u'.

Velocity gradient perpendicular to direction of flow is du/dy and

$$u' = L\frac{du}{dy}$$

Consider a portion of an eddy circulating with a velocity v' and having a cross-sectional area of flow a.

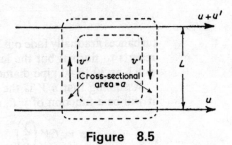

Figure 8.5

Difference of velocity between layers $= u' = 2v'$

$$v' = \tfrac{1}{2}u' = \tfrac{1}{2}L\frac{du}{dy}$$

Mass interchange from slow to fast layer per unit time is $\rho a v'$. For continuity of flow an equal mass transfers from the fast to slow layer.

Total mass interchange per unit time $= 2\rho a v'$

Change of velocity $= u'$

Rate of change of momentum $= 2\rho a v' u' = \rho a (u')^2$

This momentum transfer is occurring through the area $2a$ of both sides of the eddy and a shear stress τ is generated.

Force $=$ rate of change of momentum

$$\tau \times 2a = \rho a (u')^2$$

$$\tau = \tfrac{1}{2}\rho(u')^2 = \tfrac{1}{2}\rho \left(L\frac{du}{dy}\right)^2$$

In turbulent flow conditions locally continually vary with time and the size of eddy changes. It is usual to write

$$\text{Mean shear stress } \tau = \rho l^2 \left(\frac{du}{dy}\right)^2$$

where l is known as the *mixing length*.

In viscous flow momentum transfer between layers of fluid moving with different speeds occurs on a molecular scale and shear stress

$$\tau = \mu \frac{du}{dy}$$

where μ is the coefficient of dynamic viscosity. Comparing this with the equation

$$\tau = \rho l^2 \left(\frac{du}{dy}\right)^2$$

for turbulent flow, the equation for turbulent flow could be written

$$\tau = \varepsilon \left(\frac{du}{dy}\right)$$

where ε = eddy viscosity = $\rho l^2 \left(\frac{du}{dy}\right)$.

The coefficient ε was introduced by Boussinesq in 1877 in an attempt to explain turbulent flow, but unlike μ the value of ε is not constant.

Since $\rho l^2 du/dy$ may be taken as proportional to u the velocity at a distance y from the pipe wall, the shear stress τ at this point will, for turbulent flow, be proportional to u^2, since $\tau = \rho l^2 (du/dy)^2$, and is also a function of the velocity distribution as indicated by du/dy.

The mixing length l at a distance y from the wall can be calculated in terms of the wall shear stress τ_0 since

$$\tau = \tau_0 \left(1 - \frac{y}{R}\right) = \rho l^2 \left(\frac{du}{dy}\right)^2$$

$$l = \frac{\sqrt{\dfrac{\tau_0}{\rho}}}{\dfrac{du}{dy}} \sqrt{\left(1 - \frac{y}{R}\right)}$$

The quantity $\sqrt{\dfrac{\tau_0}{\rho}}$ is known as the *friction velocity $u*$* or *shear stress velocity* since, as shown by the following relationship involving V and f, it has the dimensions of a velocity:

$$\frac{\tau_0}{\rho V^2} = \frac{f}{2} = \text{pure number}$$

Hence $\qquad u* = \sqrt{\dfrac{\tau_0}{\rho}} = V \sqrt{\dfrac{f}{2}} = \sqrt{(mig)} = \dfrac{\sqrt{(gdi)}}{2}$

8.7 Velocity defect law

For turbulent flow in a pipe show that

$$\frac{u_m - u}{u^*} = 5 \cdot 75 \log_{10} \frac{R}{y}$$

where u_m = maximum velocity,
$\quad u$ = velocity at any distance y from the pipe wall,
$\quad R$ = pipe radius,
$\quad u^*$ = friction velocity.

Explain why this relationship does not apply close to the pipe wall.

Solution. From example **8.6**

$$\tau = \rho l^2 \left(\frac{du}{dy}\right)^2 \tag{1}$$

This equation can be solved if simple relations are assumed for the way τ and l vary with y.

Prandtl assumed that $l = ky$ where k is a constant, and if it is also assumed that $\tau = \tau_0$ = constant, equation (1) becomes

$$\tau_0 = \rho k^2 \left(y \frac{du}{dy}\right)^2$$

and $\quad u^* = \sqrt{\dfrac{\tau_0}{\rho}} = k\left(y\dfrac{du}{dy}\right) \quad$ i.e. $\quad du = \dfrac{u^*}{k}\dfrac{dy}{y}$

Integrating $\qquad u = \dfrac{u^*}{k}(\log_e y + C)$

where C = constant of integration.
Maximum velocity occurs when $y = R$ at centre.

$$u_m = \frac{u^*}{k}(\log_e R + C)$$

Thus $\qquad \dfrac{u_m - u}{u^*} = \dfrac{1}{k}\log_e \dfrac{R}{y}$

Experiments by Nikuradse indicate that $k = 0 \cdot 4$. Substituting this value and changing to common logarithms

$$\frac{u_m - u}{u^*} = 5 \cdot 75 \log_{10} \frac{R}{y} \tag{2}$$

Close to the wall the velocity of flow is very small and the flow is viscous and no longer turbulent. Equation (2) will therefore not apply.

8.8 Velocity distribution in a smooth pipe

Show that for turbulent flow in a smooth pipe

$$\frac{u}{u^*} = 5\cdot5 = 5\cdot75 \log_{10} \frac{u^*y}{v}$$

where u = velocity at radius y, u^* = friction velocity and v = kinematic viscosity.

Solution. The expression obtained for u in example **8.7** is referred to the maximum velocity u_m in the pipe and applies over the turbulent core of the flow but not in the boundary layer up to a distance y_w (Fig. 8.6) from the wall in which region flow is laminar. Therefore,

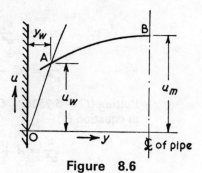

Figure 8.6

Shear stress $\tau = \mu \dfrac{du}{dy}$

Assuming a uniform velocity gradient $\dfrac{du}{dy} = \dfrac{u}{y}$ and that $\tau = \tau_0$

$$\frac{u}{y} = \frac{\tau_0}{\mu} = \frac{\tau_0}{\rho} \times \frac{\rho}{\mu} = \frac{(u^*)^2}{v} \qquad (1)$$

At the outside of the laminar sub-layer when $y = y_w$ the velocity $u = u_w$ and from equation (1)

$$\frac{u_w}{u^*} = \frac{u^*y_w}{v}$$

Experiments indicate that the thickness y_w of the laminar sub-layer is

$$y_w = 11\cdot6 \frac{v}{u^*}$$

Thus $\qquad\qquad \dfrac{u_w}{u^*} = \dfrac{u^*y_w}{v} = 11\cdot6 = \text{constant } C \qquad (2)$

From equation (1) the velocity distribution curve in the laminar sub-layer is a straight line (OA, Fig. 8.6) corresponding to

$$\frac{u}{u^*} = \frac{u^*y}{v}$$

In the central core of the pipe the velocity distribution follows the law obtained in example 8.7:

$$\frac{u_m - u}{u^*} = 5{\cdot}75 \log_{10} \frac{R}{y} \qquad (3)$$

At the point A where $y = y_w$ both expressions must give the same value for u.

From equation (2) $\qquad \dfrac{u_w}{u^*} = C$

and from equation (3) $\qquad \dfrac{u_w}{u^*} = \dfrac{u_m}{u^*} - 5{\cdot}75 \log_{10} \dfrac{R}{y_w}$

also $\qquad\qquad\qquad y_w = C\dfrac{v}{u^*}$

$$\frac{u_m}{u^*} - 5{\cdot}75 \log_{10} \frac{Ru^*}{Cv} = C$$

$$\frac{u_m}{u^*} = (C - 5{\cdot}75 \log_{10} C) + 5{\cdot}75 \log_{10} \frac{Ru^*}{v}$$

Putting $(C - 5{\cdot}75 \log_{10} C) = \text{constant} = A$ and substituting for u_m/u^* in equation (3)

$$\frac{u}{u^*} = A + 5{\cdot}75 \log_{10} \frac{Ru^*}{v} - 5{\cdot}75 \log_{10} \frac{R}{y}$$

$$\frac{u}{u^*} = A + 5{\cdot}75 \log_{10} \frac{u^*y}{v}$$

for the turbulent core of the pipe.

Experimental results by Nikuradse give a value of $A = 5{\cdot}5$ so that

$$\frac{u}{u^*} = 5{\cdot}5 + 5{\cdot}75 \log_{10} \frac{u^*y}{v}$$

Figure 8.7 shows the value of u/u^* plotted against u^*y/v for the laminar sub-layer and for the turbulent core. There is a close agreement between the theoretical and experimental values except in the transitional region from $u^*y/v = 8$ to 30.

8.9 Smooth pipe law for f

Assuming that

$$\frac{u}{u^*} = 5{\cdot}5 + 5{\cdot}75 \log_{10} \frac{u^*y}{v}$$

show that for turbulent flow in a smooth pipe

$$\frac{1}{\sqrt{(4f)}} = -0{\cdot}8 + 2 \log_{10} N_R \sqrt{(4f)}$$

where f = resistance coefficient in Darcy formula and N_R = Reynolds number based on pipe diameter and mean velocity.

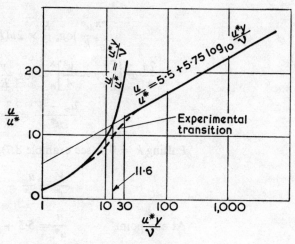

Figure 8.7

Solution. The Darcy formula for loss of head h_f due to friction is $h_f = \dfrac{4fL}{d}\dfrac{V^2}{2g}$ where L and d are the length and diameter of the pipe and V is the mean velocity.

From example **8.1** $\qquad f = 2\dfrac{\tau_0}{\rho V^2}$

Putting $\sqrt{\dfrac{\tau_0}{\rho}} = u^*$ $\qquad \dfrac{u^*}{V} = \sqrt{\dfrac{f}{2}}$

Consider an annular element of area δa at a distance y from pipe wall. The velocity in this element is u.

Discharge through element $= uda$.

$$\text{Total discharge} = \int_0^R u\,da = aV \tag{1}$$

where a = pipe area.

If the velocity had everywhere been equal to maximum velocity u_m, the discharge would have been

$$\int_0^R u_m\,da = u_m a \tag{2}$$

Subtracting equation (1) from equation (2)

$$\int_0^R (u_m - u)\,da = (u_m - V)a \tag{3}$$

From example **8.7** $\qquad u_m - u = \dfrac{u^*}{k}\log_e\dfrac{R}{y}$

and $\qquad\qquad\qquad da = 2\pi(R - y)dy.$

Substituting in (3)

$$\int_0^R \frac{u^*}{k} \log_e \frac{R}{y} \times 2\pi(R-y)dy = (u_m - V)a$$

$$\frac{2\pi}{k} u^* \left\{ \left[Ry - \frac{y^2}{4} \right]_0^R - \log_e \frac{y}{R} \left[Ry - \frac{y^2}{2} \right]_0^R \right\} = (u_m - V)\pi R^2$$

$$\frac{u_m - V}{u^*} = \frac{3}{2k} = \text{constant} = B$$

Putting $k = 0.4$ (see example 8.7), $B = 3.75$.

$$\frac{V}{u^*} = \frac{u_m}{u^*} - 3.75 \qquad\qquad (4)$$

At any point

$$\frac{u}{u^*} = 5.5 + 5.75 \log_{10} \frac{u^*y}{v}$$

Putting $u = u_m$ at centre where $y = R$

$$\frac{u_m}{u^*} = 5.5 + 5.75 \log_{10} \frac{u^*R}{v}$$

Substituting in equation (4)

$$\frac{V}{u^*} = 5.5 + 5.75 \log_{10} \frac{u^*R}{v} - 3.75$$

$$= 1.75 + 5.75 \log_{10} \frac{u^*R}{v}$$

Putting $\dfrac{V}{u^*} = \sqrt{\dfrac{2}{f}}$

$$\sqrt{\frac{2}{f}} = 1.75 + 5.75 \log_{10} \left(\frac{RV}{v} \sqrt{\frac{f}{2}} \right)$$

$$\frac{1}{\sqrt{(4f)}} = -0.91 + 2.04 \log_{10} (N_R \sqrt{(4f)})$$

where $N_R = \dfrac{Vd}{v} = \dfrac{2RV}{v}$.

The accepted "smooth pipe" law based on experimental results is

$$\frac{1}{\sqrt{(4f)}} = -0.8 + 2 \log_{10} (N_R \sqrt{(4f)})$$

This formula gives a more satisfactory value of f at high Reynolds numbers (over 100000) than the Blasius formula $f = 0.079(Vd/v)^{-1/4}$, but is more difficult to use.

8.10 Rough pipe law

Describe how the roughness of the pipe wall affects the frictional resistance to flow in pipes.

Obtain an expression for the resistance coefficient f in terms of the relative roughness for turbulent flow in pipes.

Solution. Although the flow in a pipe may be turbulent there will always be a thin laminar sub-layer adjoining the wall of the pipe on which viscous forces predominate and eddies cannot form. The roughness of the pipe wall can be measured as the average height of the projection k. If k is small compared to the thickness y_w of the laminar

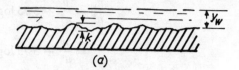

(a)

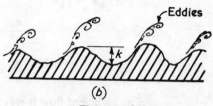

(b)

Figure 8.8

sub-layer (Fig. 8.8(a)), the roughness will be submerged in the sub-layer and will not influence the main flow so that the pipe behaves as a "smooth pipe". If k exceeds y_w the wall roughness projects into the main flow and there is a continuous generation and release of eddies (Fig. 8.8(b)) which affect the whole flow to an extent depending on the relative roughness k/D where $D =$ pipe diameter, the value of the friction coefficient f increasing as k/D increases (*see* Fig. 8.3).

The thickness of the laminar sub-layer increases as Reynolds number decreases. There will therefore be a certain Reynolds number below which a given relative roughness k/D has no influence and the value of f is given by the smooth pipe law as may be seen from the line CD of Fig. 8.3. As the Reynolds number increases and the boundary layer thickness decreases the curves break away from the smooth pipe law, starting with the roughest surface (k/D large or R/k small). At high Reynolds numbers f becomes constant depending on k/D and skin friction is proportional to the square of the velocity.

From equation (2) of example **8.7**

$$\frac{u_m - u}{u^*} = 5 \cdot 75 \log_{10} \frac{R}{y} \tag{1}$$

At the edge of the wall layer $y = y_w$ and $u = u_w$, so that

$$\frac{u_w}{u^*} = \frac{u_m}{u^*} - 5 \cdot 75 \log_{10} \frac{R}{y_w}$$

Assuming y_w depends on the roughness k, put $y_w = \alpha k$.

$$\frac{u_w}{u^*} = \frac{u_m}{u^*} + 5 \cdot 75 \log_{10} \alpha - 5 \cdot 75 \log_{10} \frac{R}{k} \tag{2}$$

From equation (1)

$$\frac{u_m}{u^*} = \frac{u}{u^*} + 5 \cdot 75 \log_{10} \frac{R}{y}$$

Substituting in equation (2)

$$\frac{u_w}{u^*} = \frac{u}{u^*} + 5 \cdot 75 \log_{10} \frac{R}{y} + 5 \cdot 75 \log_{10} \alpha - 5 \cdot 75 \log_{10} \frac{R}{k}$$

$$= \frac{u}{u^*} + 5 \cdot 75 \log_{10} \alpha - 5 \cdot 75 \log_{10} \frac{y}{k}$$

$$\frac{u}{u^*} = \left(\frac{u_w}{u^*} - 5 \cdot 75 \log_{10} \alpha \right) + 5 \cdot 75 \log_{10} \frac{y}{k}$$

Putting

$$\frac{u_w}{u^*} - 5 \cdot 75 \log_{10} \alpha = \chi \tag{3}$$

$$\frac{u}{u^*} = \chi + 5 \cdot 75 \log_{10} \frac{y}{k} \tag{4}$$

For turbulent flow it was shown in example **8.8** that

$$\frac{u}{u^*} = 5 \cdot 5 + 5 \cdot 75 \log_{10} \frac{u^* y}{\nu} \tag{5}$$

From equations (4) and (5)

$$\chi = 5 \cdot 5 + 5 \cdot 75 \log_{10} \frac{u^* k}{\nu} \tag{6}$$

or from equations (2) and (3)

$$\chi = \frac{u_m}{u^*} - 5 \cdot 57 \log_{10} \frac{R}{k}$$

Also, from equation (4), example **8.9**,

$$\frac{u_m}{u^*} = \frac{V}{u^*} + 4 \cdot 07$$

and since $\dfrac{V}{u^*} = \sqrt{\dfrac{2}{f}}$,

$$\chi = \sqrt{\frac{2}{f}} + 4 \cdot 07 - 5 \cdot 75 \log_{10} \frac{R}{k}$$

or

$$\frac{1}{\sqrt{(4f)}} - 2 \log_{10} \frac{R}{k} = \frac{(\chi - 4 \cdot 07)}{\sqrt{8}} \tag{7}$$

For smooth pipes equation (6) applies and $\dfrac{1}{\sqrt{(4f)}} - 2 \log_{10} \dfrac{R}{k}$ varies directly as $\log_{10} u^* k / \nu$. There is then a transition to rough pipe conditions for which χ is constant and experimentally it is found that for rough pipes

$$\frac{1}{\sqrt{(4f)}} - 2 \log_{10} \frac{R}{k} = 1 \cdot 74$$

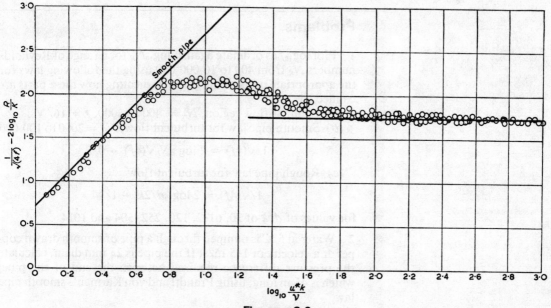

Figure 8.9

Figure 8.9 shows Nikuradse's experimental results replotted to form a single curve $\dfrac{1}{\sqrt{(4f)}} - 2\log_{10}\dfrac{R}{k}$ as ordinates and $\log_{10} u^*k/\nu$ as abscissae.

8.11 Colebrook-White formula

Show how the smooth pipe and rough pipe formula may be combined to give an approximate relation for f in terms of the relative roughness k/d and the pipe Reynolds number.

Solution. The smooth pipe formula is

$$\frac{1}{\sqrt{(4f)}} = 2\log_{10}\frac{Vd}{\nu}\sqrt{(4f)} - 0.8$$

and the rough formula is

$$\frac{1}{\sqrt{(4f)}} - 2\log_{10}\frac{d}{2k} = 1.74$$

These two formulae have been combined in the Colebrook-White formula

$$\frac{1}{\sqrt{f}} = -4\log_{10}\left(\frac{k}{3.7d} + \frac{1.255}{\dfrac{Vd}{\nu}\sqrt{f}}\right)$$

At high values of Vd/ν the second term in the bracket becomes negligible and the formula corresponds to the rough pipe law, while when k is small the first term is negligible and the formula reduces to the smooth pipe law.

Problems

1 Plot $\log_{10} f$ as ordinate against $\log_{10} N_R$ for a range of Reynolds numbers N_R from 400 to 10 000 000 using the following laws for the appropriate parts of the graph, indicating how these parts are connected.

(a) Laminar flow region, $N_R = 400$ to 3000, $f = 16/N_R$.

(b) Smooth pipe law for turbulent flow, $N_R = 2000$ to $100\ 000$

$$1/\sqrt{(4f)} = 2\log_{10} N_R\sqrt{(4f)} - 0\cdot8$$

(c) Rough pipe law for turbulent flow

$$1/\sqrt{(4f)} = 2\log_{10} d/2k + 1\cdot74$$

for values of d/k of 30, 61·2, 120, 252, 504 and 1014.

2 Water at 5°C is pumped through a pipe of smooth drawn copper at a velocity of 1·5 m/s. If the pipe is 25 mm diam, calculate the pressure difference required between the end of the pipe, which is 45 m long, using Prandtl and von Karman's smooth pipe law.
Answer 50·22 kN/m²

3 A water main 0·6 m in diam discharges 0·45 m³/s. The surface of the pipe is rough and may be taken as having protuberances of effective height 1·25 mm. Calculate the loss of head per unit length.

Calculate also the loss of head per unit length in an 18 mm diam pipe of the same material in which water at 5°C is flowing at a velocity of 1·5 m/s.
Answer 0·00507, 0·176

4 A smooth pipe 3·75 mm diam conveys 9000 m³ of water per 24 h at a temperature of 10°C. The coefficient of viscosity and the density of the water at that temperature are respectively $1\cdot309 \times 10^{-3}$ kg/m-s and 999·7 kg/m³. Calculate the Reynolds number for the flow. State whether the flow is streamline or turbulent, giving the reason for your statement. Find the head loss per kilometre of pipe with the aid of the following table.

f	0·0062	0·0050	0·00408	0·00345
$\log Re$	4·4	4·8	5·2	5·6

Answer $Re = 270\ 000$, 1·84 m/km

5 The velocity distribution in a circular pipe of radius R can be expressed as $u = u_0(y/R)^{1/m}$, where u = velocity at a point distant y from the wall and u_0 = velocity at the axis.

Demonstrate that if a pitot tube is placed at $0\cdot25R$ from the wall, the pitot registers the correct mean velocity with $\pm0\cdot5$ per cent for a range of m from 4 to 10.

Find the kinetic energy per unit-weight of flow in terms of the mean velocity, when $m = 7$.
Answer $1\cdot059V^2/2g$

6 Incompressible fluid is flowing through a pipe of circular cross-section. Calculate from first principles the radius at which a single reading of the velocity of flow will give directly the mean velocity of flow across the section when the flow is (1) wholly laminar, (2) wholly turbulent. In case (1) assume that the velocity distribution is parabolic, and in (2) that it is given by $u = u_0(y/a)^{1/m}$, where y is the distance from the wall of the pipe, u_0 is the velocity on the axis and a is the radius of the pipe.
Answer $0 \cdot 707a$, $0 \cdot 758a$

7 Indicate briefly the findings of experiments on the flow of liquids through various types of internal finish pipes at various speeds and show the variation of the Darcy coefficient of friction f with Reynolds number.

8 Define the term critical velocity used in connexion with flow through pipes and explain the reason for its having an upper and a lower value.

The following data were obtained from a test on a 3 m length of 10 mm diam smooth pipe using water whose viscosity was $0 \cdot 013$ poises, the velocity being gradually increased

Velocity (m/s)	0·183	0·213	0·244	0·274	0·305	0·335
Loss of head (m of water)	0·022	0·026	0·030	0·034	0·037	0·050

Velocity (m/s)	0·381	0·457	0·609	0·914	1·524	3·050
Loss of head (m of water)	0·083	0·147	0·295	0·600	1·470	4·950

Estimate by drawing a suitable curve, the value of the Reynolds number Re at which the critical velocity probably occurs and also show that the friction coefficient f for values of Reynolds number exceeding about 5000 is given approximately by $0 \cdot 11(Re)^{-1/4}$.
Answer 2380

9 A thin oil having a density of 800 kg/m³ is pumped through a 300 mm diameter pipeline $6 \cdot 3$ km long; the coefficient of viscosity of the oil is $0 \cdot 01865$ poises. The roughness of the inner surface of the pipe is such that the mean height of the roughness projections is $0 \cdot 75$ mm. The quantity of flow through the pipe is $0 \cdot 22$ m³/s. Using the roughness coefficient curves, calculate the head lost in friction in N/m² and the power required to drive the pump if the overall efficiency of the pump is 75 per cent.
Answer 2050 kN/m², 601 kW

10 Oil of specific gravity $0 \cdot 8$ and having a kinematic viscosity of $1 \cdot 858 \times 10^{-5}$ m²/s is pumped through a 150 mm diam pipeline of 3000 m in length. Find the power required to pump 125 m³/h.

If the oil is heated until its kinematic viscosity is $1 \cdot 858 \times 10^{-6}$ m²/s, find the power now required to pump the same quantity of oil as before.

Assume $f = 0 \cdot 08 \, Re^{-0.25}$ for turbulent flow.
Answer $30 \cdot 6$ kW, $17 \cdot 2$ kW

9

Flow round totally immersed bodies

The forces exerted between a solid body and a fluid flowing round it will be largely of two kinds, firstly the frictional forces due to viscosity which are generated in the boundary layer adjacent to the solid surface (already discussed in Volume 1 Chapter 14), and secondly those generated in the region outside the boundary layer where the velocity and pressure in the free stream are affected by the physical presence of the body and its associated boundary layer and wake.

9.1 Boundary layer separation and wake formation for flow over a convex curved surface

Describe boundary layer separation and subsequent wake formation for flow over the curved convex surface of a body immersed in a moving fluid.

Solution When a viscous fluid flows over a solid surface a boundary layer is formed in which the fluid velocity changes from zero at the solid surface to the free stream velocity at the boundary layer edge (see Volume 1 Chapter 14). When a fluid passes over the convex surface of a solid, such as a cylinder, the boundary layer will tend to separate from this surface just aft of the point of maximum thickness where the surface curvature requires that the fluid should decelerate. As shown in Fig. 9.1 the fluid velocity in the boundary layer is increasing in the

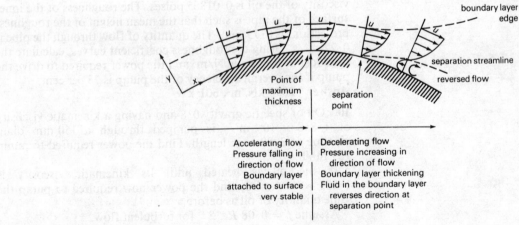

Accelerating flow
Pressure falling in direction of flow
Boundary layer attached to surface very stable

Decelerating flow
Pressure increasing in direction of flow
Boundary layer thickening
Fluid in the boundary layer reverses direction at separation point

Figure   9.1

region upstream of this maximum thickness point and then decreasing downstream of it. Where the flow is accelerating the boundary layer thickness reduces slightly but starts increasing rapidly when the flow begins to decelerate.

While the fluid outside the boundary layer is accelerating the pressure gradient dp/dx is said to be negative, or "favourable", and can lead to a reduction in the boundary layer thickness.

At, or about, the point of maximum thickness however the stream-wise pressure gradient dp/dx reduces to zero and downstream of this point the pressure gradient becomes positive, or "adverse".

Adverse pressure gradients oppose the flow, reducing the velocities in the boundary layer and increasing its thickness. These two effects combine to reduce the velocity gradient du/dy at the wall. At the separation point du/dy is zero. Thereafter a reversed flow region occurs producing vortices and subsequent wake formation.

The near wake, immediately downstream of the solid body, is in general a region where the static pressure and/or the flow velocity are lower than in the undisturbed stream. Near the solid body the static pressure in the wake is close to that at the separation point. This low pressure leads to a pressure difference between the high pressures over the upstream surfaces and the low pressures on the downstream sur-faces in the near wake thus causing a pressure drag force on the body in the direction of fluid flow.

The structure of the wake depends on the Reynolds number of the flow and the detailed shape of the body. Immediately aft of the body, following boundary layer separation, strong vortices will be formed which may detach themselves at regular intervals of time forming the von Karmann *vortex street*.

In the far wake further downstream the static pressure rises to the undisturbed value but the fluid velocity will remain below the un-disturbed value for a great distance until the effects of viscosity have obliterated the wake. Fig. 9.2 shows some of the characteristics of a wake downstream of a prismatic body.

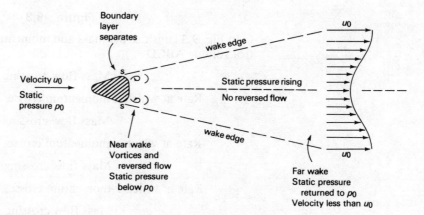

Figure 9.2

9.2 Drag force on a prismatic solid

(a) Apply the momentum equation to obtain an integral expression for the fluid drag coefficient of a prismatic body which produces a symmetrical far wake of total width $2h$ when immersed in a fluid of density ρ_0 moving with a velocity u_0.

(b) A far wake velocity profile downstream of a prismatic solid is 20 cm wide and has the approximate form

$$\frac{u}{u_0} = 0\cdot6 + 0\cdot4\,\frac{y}{h}$$

Calculate the drag coefficient if the thickness of the prism normal to the flow is 6 cm.

Solution (a) In Fig. 9.2 the far wake velocity profile shows that the fluid suffers a momentum deficit in passing over the solid body. The net force on the fluid causing this rate of change of momentum must be equal and opposite to the force exerted on the body by the fluid. Thus given the velocity profile in the far wake the momentum equation can be applied to calculate the drag force.

Since in the far wake the static pressure has returned to its undisturbed value, the only force acting on the fluid in the x-direction is due to the prism.

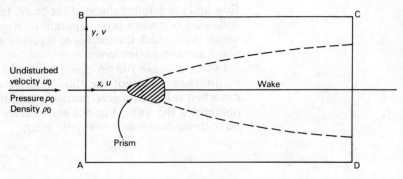

Figure 9.3

In Fig. 9.3 consider the mass and momentum flows across the control volume ABCD

$$\text{Mass flow crossing AB} = \int_A^B \rho_0 u_0\,dy$$

$$\text{Rate at which } x\text{-momentum crosses AB} = \int_A^B \rho_0 u_0^2\,dy$$

$$\text{Mass flow crossing BC} = \int_B^C \rho_0 v\,dx$$

$$\text{Rate at which } x\text{-momentum crosses BC} = \int_B^C \rho_0 v u_0\,dx$$

$$\text{Mass flow crossing CD} = \int_C^D \rho_0 u\,dy$$

$$\text{Rate at which } x\text{-momentum crosses CD} = \int_C^D \rho_0 u^2\,dy$$

$$\text{Mass flow crossing AD} = \int_D^A \rho_0 v\,dx$$

$$\text{Rate at which } x\text{-momentum crosses AD} = \int_D^A \rho_0 v u_0\,dx$$

Thus

Total rate of change of x-momentum through ABCD
= momentum leaving through BC and CD less momentum entering through AD and AB

$$= \int_B^C \rho_0 v u_0 dx + \int_C^D \rho_0 u^2 dy - \int_D^A \rho_0 v u_0 dx - \int_A^B \rho_0 u_0^2 dy$$

$$= u_0 \left\{ \int_B^C \rho_0 v dx - \int_D^A \rho_0 v dx \right\} + \left\{ \int_C^D \rho_0 u^2 dy - \int_A^B \rho_0 u_0^2 dy \right\}$$

(1)

For continuity of flow

Mass flow crossing AB and AD = Mass flow crossing CD and BC

$$\int_A^B \rho_0 u_0 dy + \int_D^A \rho_0 v dx = \int_C^D \rho_0 u dy + \int_B^C \rho_0 v dx$$

$$\int_B^C \rho_0 v dx - \int_D^A \rho_0 v dx = \int_A^B \rho_0 u_0 dy - \int_C^D \rho_0 u dy$$

(2)

Force F_x exerted on the fluid by the body in the x-direction = rate of change of momentum in the x-direction

Thus combining equations (1) and (2)

$$F_x = u_0 \left\{ \int_A^B \rho_0 u_0 dy - \int_C^D \rho_0 u dy \right\} + \int_C^D \rho_0 u^2 dy - \int_A^B \rho_0 u_0^2 dy$$

$$= \int_C^D \rho_0 u^2 dy - \int_C^D \rho_0 u_0 u dy$$

The drag force D on the prism is equal and opposite to the force on the fluid

$$D = \int_C^D \rho_0 u(u_0 - u)\, dy$$

Outside the limits of the wake $u_0 - u = 0$ and if the wake width is $2h$

$$\text{Drag } D = \int_{-h}^{+h} \rho_0 u(u_0 - u)\, dy$$

$$\text{Drag coefficient } C_D = \frac{\text{Drag force}}{\frac{1}{2}\rho_0 u_0^2 \times \text{projected frontal area}}$$

If the projected frontal area for unit span is $t \times 1$

$$C_D = \frac{2}{t} \int_{-h}^{+h} \frac{u}{u_0} \left(1 - \frac{u}{u_0} \right) dy$$

$$= \frac{2\theta}{t}$$

where θ is the momentum thickness of the wake.

(b) As shown above $C_D = 2\theta/t$ where

$$\theta = \int_{-h}^{+h} \frac{u}{u_0} \left(1 - \frac{u}{u_0} \right) dy$$

Putting $y/h = \eta$ then $dy = h\, d\eta$ and the limits of integration are changed since when $y = -h$, $\eta = -1$ and when $y = +h$, $\eta = +1$ giving

$$\theta = h \int_{-1}^{+1} \frac{u}{u_0} \left(1 - \frac{u}{u_0} \right) d\eta$$

But from the question $\dfrac{u}{u_0} = 0\cdot6 + 0\cdot4y/h$

$$\therefore \theta = h\int_{-1}^{+1} (0\cdot6 + 0\cdot4\eta)(1 - 0\cdot6 + 0\cdot4\eta)d\eta$$

$$= h\int_{-1}^{+1} (0\cdot24 - 0\cdot08\eta - 0\cdot16\eta^2)d\eta$$

$$= h\left[0\cdot24\eta - 0\cdot04\eta^2 - \frac{0\cdot16}{3}\eta^3\right]_{-1}^{+1}$$

$$= h\left\{\left(0\cdot24 - 0\cdot04 - \frac{0\cdot16}{3}\right) - \left(-0\cdot24 - 0\cdot04 + \frac{0\cdot16}{3}\right)\right\}$$

$$= 0\cdot587h$$

where $h = \tfrac{1}{2} \times$ width of wake $= 0\cdot1$ m
$$\therefore \theta = 0\cdot0587 \text{ m}$$

and $C_D = \dfrac{1 \times 0\cdot0587}{t} = \dfrac{0\cdot1174}{0\cdot06} = \mathbf{1\cdot957}$

9.3 Drag on bluff and streamline bodies

> Define skin friction drag, pressure drag and total drag and distinguish between streamline and bluff bodies in terms of their relative magnitudes.

Solution When a solid body is immersed in a stream of fluid boundary layers are formed on those surfaces facing upstream and, in general, separate forming a wake flow downstream.

Skin friction drag is due to viscous shear forces produced at the body surface predominantly in those regions to which the boundary layer is attached. On any element of area dA,

Component of shear force in flow direction $= \tau_w\cos\theta.dA$

where

τ_w is the local viscous shear stress at the body surface
θ is the inclination of the area dA to the flow direction

Integrating over the whole surface

Skin friction drag $D_F = \oint \tau_w\cos\theta.dA$

For surfaces normal to the direction of flow $\theta = 90°$ and the skin friction drag force will therefore be zero.

Pressure drag, sometimes called *form drag*, is due to the unbalanced pressures which exist between the relatively high pressures on the upstream body surfaces and the lower pressures on the downstream surfaces. On any area dA

Component of pressure force in flow direction $= P_L\sin\theta, dA$

where

P_L is the local static pressure acting on the body surface
θ is the inclination of dA to the flow direction

Integrating over the whole surface

$$\text{Pressure drag } D_P = \oint P_L \sin\theta, \, dA$$

Pressures acting on a surface in line with the flow direction for which $\theta = 0$ will not contribute to the pressure drag.

The *total drag* on a body, sometimes called the *profile drag*, is the sum of the skin friction drag and the pressure drag

$$D_T = D_F + D_P$$

The relative magnitudes of the skin friction drag and the pressure drag can be used to draw a distinction between streamlined and bluff bodies.

A *streamlined* body is invariably elongated in the flow direction with the boundary layer attached over the major part of its surface. A high proportion of the total drag is due to skin friction drag. The ultimate streamlined body would be a thin flat plate aligned with the flow direction for which the skin friction drag would constitute the total drag on the body.

A *bluff* body has a short overall length in the streamwise direction relative to its thickness. Flow separation takes place at, or about, the point of maximum thickness resulting in a large pressure drag on the body and a small skin friction drag. A thin flat plate normal to the stream would have maximum bluffness since none of the shear stress forces can act in the flow direction and the total drag is due solely to the pressure drag.

9.4 Drag force on rectangular flat plate

Define the pressure coefficient C_P and calculate the drag force on a rectangular flat plate of $0 \cdot 5$ m span and $0 \cdot 05$ m chord when it is held

(i) at zero incidence to a stream of air of velocity 30 m/s, viscosity $0 \cdot 000018$ kg/m-s and density $1 \cdot 2$ kg/m³.

(ii) At 90° incidence assuming average C_P over the upstream face of $+0 \cdot 85$ and an average C_P over the downstream face of $-1 \cdot 6$.

Solution The static pressure distribution over the surface of a body can be expressed in terms of a *pressure coefficient* C_P

$$C_P = \frac{p_L - p_0}{\frac{1}{2}\rho_0 u_0^2}$$

where

p_L is the local static pressure on the body surface
p_0 is the static pressure in the undisturbed fluid stream
$\frac{1}{2}\rho_0 u_0^2$ is the dynamic pressure of the indisturbed fluid stream

If the total pressure P_0 is the same throughout the flow field

$$P_0 = p_0 + \tfrac{1}{2}\rho_0 u_0^2 = p_L + \tfrac{1}{2}\rho_0 u_L^2$$

where u_L is the local fluid velocity

$$p_L - p_0 = \tfrac{1}{2}\rho_0(u_0^2 - u_L^2)$$

so that

$$C_P = 1 - \frac{u_L^2}{u_0^2}$$

At the stagnation point $C_P = +1\cdot0$ while in regions where the local velocity exceeds the undisturbed value C_P will become negative and may be referred to as areas of suction.

(i) To calculate the drag on a flat plate held parallel to the stream direction a simple boundary layer calculation is sufficient, see Volume 1 Chapter 14, example 14.4 from which

$$\text{Drag coefficient } C_D = 1\cdot155(Re_l)^{-1/2}$$

If $u_0 = 30$ m/s, $\rho_0 = 1\cdot2$ kg/m³, $\mu_0 = 1\cdot8 \times 10^{-5}$ kg/m-s the Reynolds number at the trailing edge of the plate is

$$Re_l = \frac{1\cdot2 \times 30 \times 0\cdot05}{0\cdot000018} = 10^5$$

Assuming a laminar boundary layer on both sides of the plate

$$C_D = 1\cdot155\,(10^5)^{-1/2} = 0\cdot003165$$
$$\text{Skin friction drag } D_F = C_D \times \text{wetted area}$$
$$= 0\cdot003165\,(2 \times 0\cdot05 \times 0\cdot5)$$
$$= 1\cdot58 \times 10^{-4}\text{N}$$

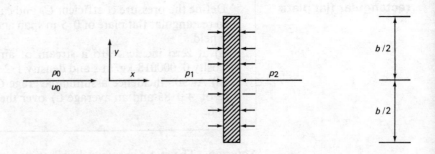

Figure 9.4

(ii) If the plate is held normal to the flow the drag will be solely due to unbalanced pressure forces as shown in Fig. 9.4. The force on an element of area dA due to the pressure difference between the upstream and downstream surfaces will be

$$dD_P = (p_1 - p_2)dA$$

If the flow over the plate is assumed two-dimensional

$$dA = s.dy$$

where s is the span measured perpendicular to the diagram

therefore

$$dD_P = (p_1 - p_2)s\,dy$$

and since

$$p_1 - p_2 = (p_1 - p_0) - (p_2 - p_0)$$
$$dD_P = \tfrac{1}{2}\rho_0 u_0^2 (C_{P1} - C_{P2})s\,dy$$

where C_{P1} and C_{P2} are the local pressure coefficients on the upstream and downstream surfaces respectively.

The total drag D_P is obtained by integrating from $y = -b/2$ to $+b/2$ where b is the chord width.

$$D_P = \tfrac{1}{2}\rho_0 u_0^2 \int_{-b/2}^{+b/2} (C_{P1} - C_{P2})s\,dy$$
$$= \tfrac{1}{2}\rho_0 u_0^2 (C_{P1} - C_{P2})s\,b$$

Substituting the figures in the question

$$D_P = \tfrac{1}{2} \times 1\cdot 2 \times 30^2 \{0\cdot 85 - (-1\cdot 6)\} \times 0\cdot 5 \times 0\cdot 05 \text{ N}$$
$$= 33\cdot 07 \text{ N}$$

9.5 Pressure and drag coefficients for flow round a cylinder

With the aid of diagrams describe the flow patterns around a circular section cylinder immersed in a moving fluid for a range of Reynolds numbers from 10^{-1} to 10^6. Show how the pressure coefficient distribution around the cylinder and its drag coefficient vary over this range of Reynolds numbers.

Discuss the structure of the wake for the above flow fields, giving an account of the vortex shedding characteristics.

Solution The flow field around a cylinder at Reynolds numbers from 10^{-1} to 10^6 is of particular interest because it demonstrates features which occur in the flow fields about other bodies immersed in a moving fluid. For a cylinder the Reynolds number is defined as $\rho_0 u_0 D/\mu_0$ where D is the diameter of the cylinder.

(i) $Re = 10^{-1}$ *or less.* Inertia forces are negligible and the flow remains attached over the entire surface. For a Reynolds number of $0\cdot 1$ and a cylinder of $0\cdot 01$ m diameter in a stream of water of density $\rho = 1000$ kg/m³ and viscosity $\mu_0 = 0\cdot 002$ kg/m-s the approach velocity would be

$$u_0 = \frac{Re \times \mu_0}{\rho_0 \times D} = \frac{0\cdot 1 \times 0\cdot 002}{1000 \times 0\cdot 01} \text{ m/s}$$

$$= 20 \ \mu\text{m/s} \quad \text{or } 0\cdot 072 \text{ m/hr}$$

This is known as *creeping flow* and will be laminar everywhere. Such flows may occur in, for example, water seepage through a porous medium around a pipe. Low Reynolds numbers also occur where the diameter of the cylinder is very small. Thus for water flowing at $0 \cdot 1$ m/s the Reynolds number will be $0 \cdot 1$ when the cylinder diameter is 2μm.

The flow field and the C_P distribution at this Reynolds number agrees closely with that predicted theoretically for an ideal fluid which has zero viscosity (see Chapter 4), and are shown in Fig. 9.5. The flow field is symmetrical on either side and upstream and downstream and C_P has a value of $+1$ at both the forward and rear stagnation points.

The drag coefficient C_D is about 50 at this Reynolds number and the drag force is entirely due to skin friction.

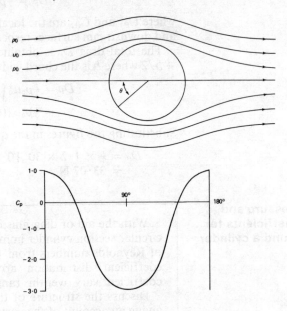

Figure 9.5 Re < 0.1

(ii) *Re = 1 to 50*. As the Reynolds number increases the drag coefficient falls until, in the Reynolds number range $1 < Re < 50$, separation occurs near the rear stagnation point. Two symmetrical vortices are formed which remain attached to the cylinder as shown in Fig. 9.6.

At a Reynolds number of 10 the drag coefficient drops to 5. The flow is still completely laminar including the vortex flows. To give an idea of scale, a Reynolds number of 10 corresponds to the flow of oil $\rho_0 = 850$ kg/m³ $\mu_0 = 0 \cdot 004$ kg/m-s at $0 \cdot 1$ m/s around a cylinder $0 \cdot 47$ mm in diameter.

(iii) *Re = 50 to 10³*. Increasing the Reynolds number further causes the vortices to elongate in the streamwise direction and eventually to detach themselves alternately from either side of the cylinder at a Reynolds number of about 100. The drag coefficient is about $1 \cdot 5$, although this will represent the average of the unsteady loads induced by vortex shedding. The pressure coefficient distribution will also fluctuate, particularly near the upper and lower separation points. Fig. 9.7

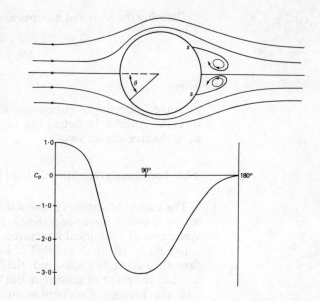

Figure 9.6 Re 1 to 50

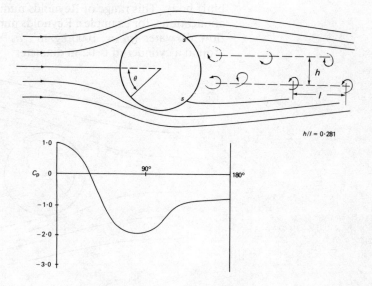

h/l = 0·281

Figure 9.7 Re = 100

shows the flow field and average C_P distribution at this Reynolds number. The low values of C_P over the downstream surfaces has been explained in Example 9.4.

The flow field in Fig. 9.7 comprises a laminar boundary layer over the surface of the cylinder followed by vortex shedding and a wake which becomes turbulent. The frequency of vortex shedding is expressed by the *Strouhal number Str* which for a cylinder is a weak function of Reynolds number.

Defining the Strouhal number as

$$Str = \frac{fd}{u_0}$$

where

f is the frequency of vortex shedding in Hz
d is the cylinder diameter in m
u_0 is the free stream velocity

it has been found that $Str = 0 \cdot 198 \left(1 - \dfrac{19 \cdot 7}{Re}\right)$ for $250 < Re < 2 \times 10^5$

The stream of vortices in the wake of a cylinder is known as the *von Karman vortex street* and the unsteady loads produced can lead to vibrations of cylindrical structures.

(iv) $Re = 10^3$ to 2×10^5. For Reynolds numbers in this range the flow field remains unchanged, the boundary layer separates just forward of the point of maximum thickness. Fig 9.8 shows the flow field and the average C_P distribution. The drag coefficient is largely independent of Reynolds number over this range at approximately $1 \cdot 2$. The drag is now almost entirely pressure drag, the cylinder acting as a bluff body. This range of Reynolds number is common in engineering applications, for example a Reynolds number of 10^3 corresponds to the flow of water ($\rho_0 = 1000$ kg/m³, $\mu_0 = 0 \cdot 002$ kg/m-s) at $0 \cdot 1$ m/s around a cylinder of $0 \cdot 02$ m diameter.

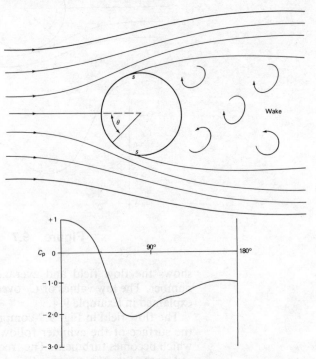

Figure 9.8 Re 10^3 to 2×10^5

(v) *Re = 2 × 10⁵*. At or about this Reynolds number the boundary layer on the surface of the cylinder becomes turbulent and more resistant to separation, causing the separation points to move aft of the point of maximum thickness, producing a much narrower wake, reduction of the value of C_D to about $0·3$ and the elimination of the vortex street. Vortices continue to be shed but in a random fashion which in general do not induce vibration of cylindrical structures. Fig. 9.9 shows the flow field and C_P distribution.

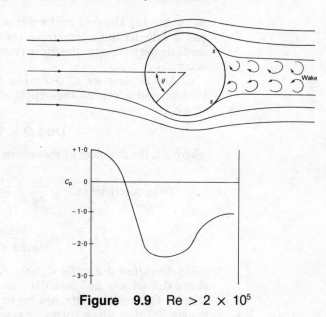

Figure 9.9 Re > 2 × 10⁵

Fig. 9.10 shows the variation of the drag coefficient with Reynolds number for a cylinder, having the turbulent boundary layer and narrow wake flow field described above.

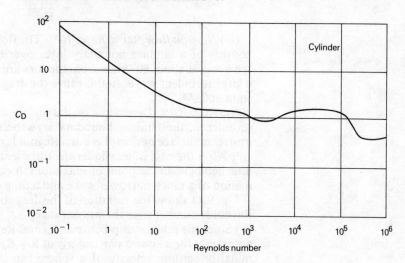

Figure 9.10

9.6 Flow past a sphere

(*a*) Discuss the flow past a sphere for the Reynolds number range from $0 \cdot 01$ up to 10^5. Explain how the relationship between the drag coefficient C_D and Reynolds number can be approximated for the different flow regimes.

(*b*) Discuss how the concept of terminal velocity for a sphere falling under gravity can be used to determine fluid viscosity.

Solution (*a*) The flow past a sphere resembles in many ways the flow past a cylinder but alternate vortex shedding cannot occur and drag loads on the sphere are therefore constant. Four different flow regimes occur.

(i) *Stokes flow Re* $\leqslant$ *0·2* The flow is completely attached and laminar and a proven theoretical solution has been obtained from which

$$\text{Drag } D = 3\pi d\mu_0 u_0$$

where d is the diameter of the sphere giving

$$\text{Drag coefficient } C_D = \frac{D}{\frac{1}{2}\rho_0 u_0^2 \times \frac{1}{4}\pi d^2}$$

$$= \frac{3\pi d\mu_0 u_0}{\frac{1}{2}\rho_0 u_0^2 \times \frac{1}{4}\pi d^2} = \frac{24\mu_0}{\rho_0 u_0 d} = \frac{24}{Re}$$

(ii) *Allen flow 0·2 < Re < 500* As the Reynolds number increases above $0 \cdot 2$ an attached toroidal vortex forms around the rear of the sphere. This remains attached up to a Reynolds number of approximately 200 after which vortex rings are shed in a random fashion. The relationship between sphere drag coefficient and Reynolds number in this range is given by

$$C_D = \frac{18 \cdot 5}{Re^{0 \cdot 6}}$$

(iii) *Newton flow 500 < Re < 10⁵* The flow field around the sphere consists of a laminar boundary layer over the upstream face which separates at about 80° back from the forward stagnation point causing a large turbulent wake. In this range the drag coefficient is nearly constant at $0 \cdot 44$.

(iv) *Re > 10⁵* Depending on the surface roughness and free stream turbulence, the laminar boundary layer becomes turbulent over the upstream surface of the sphere at a Reynolds number of approximately 10^5. As in the case of a cylinder the turbulent boundary layer remains attached beyond the point of maximum thickness resulting in the formation of a much narrower wake and a drag coefficient of $0 \cdot 3$.

Fig. 9.11 shows the variation of the drag coefficient of a sphere with Reynolds number over the whole range.

(*b*) Since the relationship between C_D and Reynolds number is known accurately for a sphere particularly at low Reynolds numbers, the terminal or settling velocity of a sphere can be predicted and used to calculate the viscosity of the fluid.

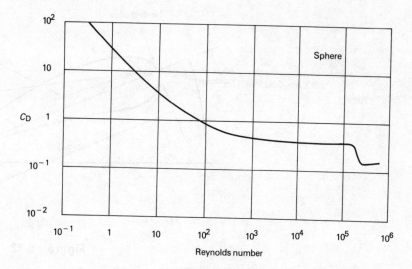

Figure   9.11

Equating the weight of the sphere to the sum of the drag force and the buoyancy force

$$\rho_P g \frac{\pi d^3}{6} = 3\pi d \mu_0 u_T + \rho_0 g \frac{\pi d^3}{6}$$

where

ρ_P is the density of the material of the sphere
u_T is the terminal or settling velocity of the sphere

$$U_T = \frac{g d^2}{18 \mu_0} (\rho_P - \rho_0)$$

In a falling sphere viscometer the diameter and weight of the sphere are chosen so that, at the terminal velocity Reynolds number is less than $0 \cdot 2$, Stokes flow occurs and the above equations hold.

9.7 Lift and drag on an aerofoil

> (a) Explain with the aid of diagrams the flow over a symmetrical aerofoil explaining how lift forces are generated. Show typical lift and drag characteristics for an aerofoil section.
> (b) Describe briefly how non-symmetrical aerofoil blades are used in rotodynamic pumps and turbines.

Solution (a) An aerofoil is a streamlined body designed to produce lift with a minimum drag. Fig. 9.12 shows a typical aerofoil and some of the terms relating to it. If the aerofoil is symmetric the camber line coincides with the chord line as in Fig. 9.13 which shows the flow field on such an aerofoil aligned with the flow at *zero incidence*.

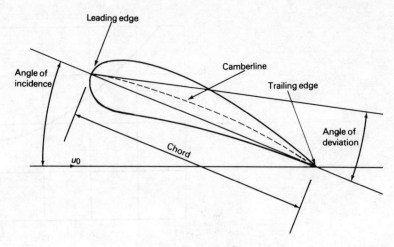

Figure 9.12

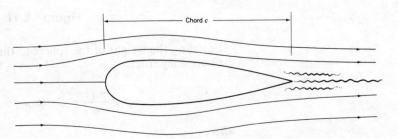

Figure 9.13

The streamlines converge near the point of maximum thickness as the fluid accelerates around the section. The increase in velocity of the fluid is accompanied by a reduction in static pressure. The pressure coefficient C_P on the top and bottom surfaces becomes negative. Downstream of the maximum thickness point the flow decelerates the pressure rises and eventually boundary layer separation will occur towards the trailing edge of the aerofoil. A small turbulent wake is usually formed. The drag coefficient at zero incidence will be between $0\cdot01$ and $0\cdot03$ for a typical aerofoil section where

$$C_D = \frac{\text{Drag}}{\frac{1}{2}\rho_0 u_0^2 cs}$$

with c = chord and s = span of the aerofoil.

The C_P distribution over the upper and lower surfaces will be identical as shown in Fig. 9.14. There is no lift force.

When the aerofoil section is presented to the oncoming fluid at an angle of incidence a lift force is generated which can be defined in terms of a lift coefficient C_L so that Lift $L = \frac{1}{2}C_L\rho_0 u_0^2 A$. The flow field is modified as shown in Fig. 9.15. The forward stagnation point moves downwards slightly, the boundary layer separation point on the upper

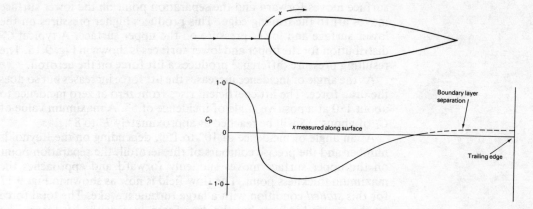

Figure 9.14

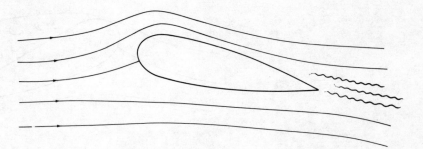

Figure 9.15

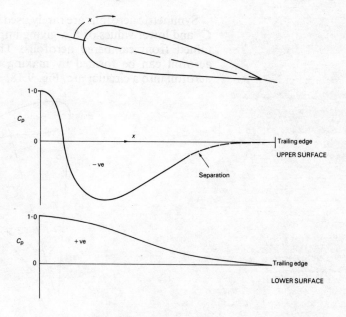

Figure 9.16

surface moves forward and the separation point on the lower surface moves aft to the trailing edge. This produces higher pressures on the lower surface and lower pressure on the upper surface. A typical C_P distribution for the upper and lower surfaces is shown in Fig. 9.16. The resulting pressure difference produces a lift force on the aerofoil.

As the angle of incidence increases the lift force increases but so does the drag force. The lift coefficient rises from zero at zero incidence to about $1 \cdot 0$ at a positive angle of incidence of $5°$. A maximum value of C_L of about $1 \cdot 5$ will be reached at approximately $7°$ to $8°$.

At an angle of incidence of $10°$ to $15°$, depending on the Reynolds number and the precise contours of the aerofoil, the separation point on the upper surface moves suddenly forward and approaches the maximum thickness point. The flow field is now as shown in Fig. 9.17 for this *stalled* condition with a large turbulent wake. The total force on the aerofoil is large but the drag force exceeds the lift force. The drag coefficient rises towards $0 \cdot 5$ and the lift coefficient will be dropping below $0 \cdot 5$.

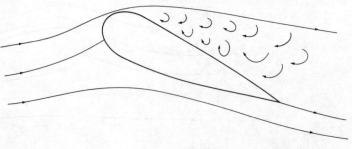

Figure   9.17

Symmetric aerofoils are rarely used in practice since higher values of C_L and lower values of C_D, giving improved C_L/C_D ratios, can be obtained from cambered aerofoils. The simplest form of cambered aerofoil can be formed by making the centre line of a symmetric aerofoil into a circular arc, Fig. 9.18. The curvature of the camber line

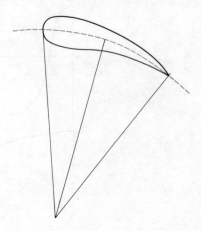

Figure   9.18

and the asymmetric distribution of aerofoil thickness about this line can be combined to give the required aerodynamic characteristics for any particular application. For aircraft, high lift at low speed can be obtained from relatively thick aerofoils which have good stalling characteristics. For high speed aircraft slender asymmetric aerofoils are used having an almost straight camber line.

The lift and drag characteristics for a typical symmetric aerofoil are shown in Fig. 9.19.

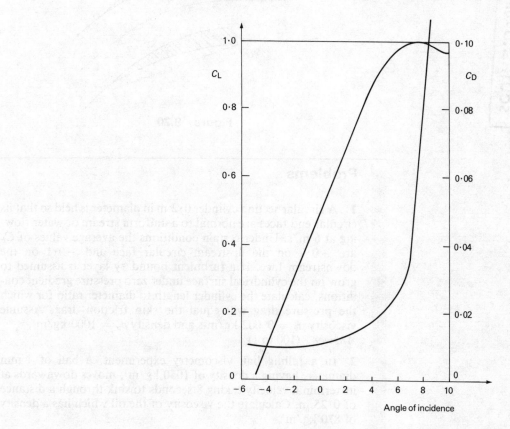

Figure 9.19

(*b*) Blades of aerofoil section can be used in rotodynamic machinery, the object being to achieve a low drag coefficient and a high lift coefficient thus reducing losses and improving the energy transfer between the blades and the fluid. Aerofoil blades often form a cascade, the individual blades being sufficiently close together as shown in Fig. 9.20 for the characteristics of each aerofoil to be enhanced by the support of its neighbours. The ratio of the blade chord *c* to the space *s* between the blades *c/s* is called the *blade solidity*. In Fig 9.20 with *c/s* = 1·0 the fluid is restrained from separating from the upper surface of each blade by the guiding effect of the lower surface of the adjacent blade.

Separation and stall still occur in a cascade at a high angle of incidence but the flow becomes three dimensional with streamwise vortices generated by the pressure gradient between the aerofoil sections.

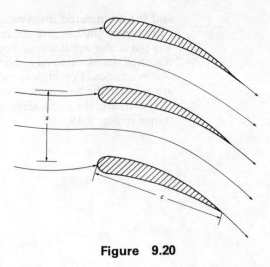

Figure 9.20

Problems

1 A circular section cylinder $0 \cdot 2$ m in diameter is held so that its circular end faces are normal to a uniform stream of water flowing at 6 m/s. Under certain conditions the average values of C_P are $+0 \cdot 7$ on the upstream circular face and $-0 \cdot 1$ on the downstream face. If a turbulent boundary layer is assumed to grow on the cylindrical surface under zero pressure gradient conditions, calculate the cylinder length to diameter ratio for which the pressure drag will equal the skin friction drag. Assume viscosity $\mu_0 = 0 \cdot 002$ kg/m-s and density $\rho_0 = 1000$ kg/m³.
Answer (100 to 1)

2 In a falling ball viscometry experiment, a ball of 1 mm diameter, having a density of 1050 kg/m³, moves downwards at its terminal velocity taking 8 seconds to sink through a distance of $0 \cdot 25$ m. Calculate the viscosity of the oil which has a density of 870 kg/m³.
Answer $0 \cdot 0094$ kg/m-s

3 A circular chimney is 1 m diameter and 30 m high. Calculate (i) the minimum wind velocity which will just cause vortex shedding to occur and (ii) the frequency of shedding.
What is the bending moment at the chimney base when a wind of 20 m/s is blowing. Air density $\rho_0 = 1 \cdot 18$ kg/m³ viscosity $\mu_0 = 1 \cdot 3 \times 10^{-5}$ Pa s.
Answer $1 \cdot 1$ m/s, $0 \cdot 218$ Hz, $84 \cdot 96$ kN m

4 In a gravel washing and sorting plant gravel of density 3400 kg/m³ is introduced into a stream of water moving upwards at $0 \cdot 5$ m/s. If the gravel particles are assumed to be approximately spherical and to have a drag coefficient of about $0 \cdot 5$, what would be the diameter of the largest particle which would be carried upwards by the water.

If the water velocity were to be doubled what would be the diameter of the smallest particle which would move downwards.

Density of water $\rho_0 = 1000$ kg/m³

Answer 4 mm, 16 mm

5 The velocity profile of the far wake behind a body of revolution approximates to the form

$$\frac{u}{u_0} = 0\cdot7 + 0\cdot3\,\frac{r}{R}$$

where R is the radius of the wake, u_0 is the free stream velocity, and u is the velocity at radius r.

If the static pressure in the wake is equal to the undisturbed value, calculate the drag coefficient of the body given that the diameter of the body at maximum thickness is $0\cdot1$ m, the radius of the wake is $0\cdot15$ m and the density of the fluid is $1\cdot2$ kg/m³.

Answer 1·6

6 A small aircraft weighing 10 000 N is flying at 60 m/s. Its wings have an effective span of 7 m and a chord of $1\cdot1$ m. Using the aerofoil data given in Fig. 9.21, calculate the power absorbed in overcoming the drag on the wings. Density of air $\rho_0 = 1\cdot18$ kg/m³.

Answer 29·44 kW

7 A free-fall parachutist and his equipment might be considered to be equivalent aerodynamically to a sphere representing the head and crash helmet and the following cylindrical members: two arms $0\cdot1$ m diameter and $0\cdot6$ m long, two legs $0\cdot15$ m diameter and $0\cdot8$ m long and a torso $0\cdot35$ m diameter and $0\cdot55$ m long.

By making suitable assumptions calculate the terminal velocity of the parachutist when falling through air of density $1\cdot2$ kg/m³ and viscosity $1\cdot8 \times 10^{-5}$ kg/m-s. The parachutist has a mean density of 1060 kg/m³.

Answer 53 m/s

8 During an investigation of aerodynamic forces on prisms, data has been obtained for pressure coefficient distribution over the four faces of a square section prism 75 mm × 75 mm held normal to the flow but rotated through 30° as shown in Fig. 9.22. The pressure coefficient distribution on each of the four faces is shown in Table 9.1.

Working from first principles calculate the pressure drag coefficient of the prism for unit width.

Answer 1·86

9 During the investigation of the prism in Problem 8, a pitot traverse of the wake well downstream of the model yields the assumed symmetrical wake velocity distribution given in Table 9.2.

Working from first principles calculate the total drag coefficient per unit width of the prism.

Answer 2·428

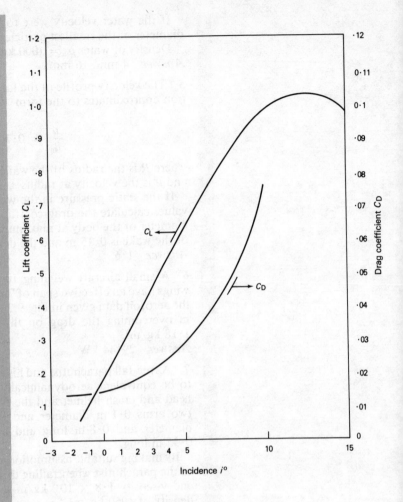

Figure 9.21

Table 9.1

S/C	C_{p1}	C_{p2}	C_{p3}	C_{Cp4}
0	−1·91	−2·0	−1·6	− ·72
·03	− ·91			− ·58
·09	− ·4			− ·6
·17	+ ·02			− ·44
·26	+ ·26			− ·17
·37	+ ·46			− ·08
·50	+ ·63			− ·20
·63	+ ·76			− ·38
·74	+ ·88			− ·53
·83	+ ·96			− ·69
·91	+ ·97			− ·90
·97	+ ·89			−1·20
1·00	− ·01			−1·72

Table 9.2

y(m)	u_2(m/s)
·24	22
·20	22
·16	20·8
·12	17·4
·08	12·4
·04	9·6
0	8·9

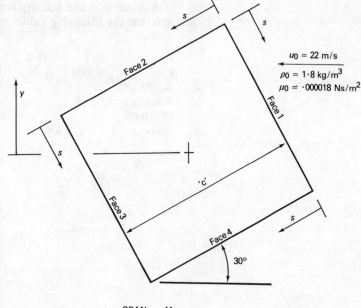

Face 2

Face 1

Face 3

Face 4

$u_0 = 22$ m/s
$\rho_0 = 1\cdot8$ kg/m^3
$\mu_0 = \cdot000018$ Ns/m^2

y

'c'

s

$30°$

SPAN = $\cdot44$ m

Figure 9.22

10 As part of an investigation into forces caused by vortex shedding from cylindrical insertion flow meters it was assumed that, at the instant of vortex shedding, the C_P distribution around the cylinder was given by $1 - 4\sin^2\theta$ for $\frac{1}{2}\pi < \theta < 2\pi$ and -3 for $0 < \theta < \frac{1}{2}\pi$.

Calculate the side force lift coefficient for the C_P given.

Answer $\dfrac{7}{6}$

11 An aerofoil having a chord of $0\cdot2$ m and a span of $1\cdot2$ m is tested in a compressed air wind tunnel at a Reynolds number of 9×10^6. The pressure and temperature of the air at the working section are 20 bar and 27°C respectively and the kinematic viscosity is $7\cdot9 \times 10^{-7}$ m^2/s.

The aerodynamic force acting on the aerofoil is measured by a 2-component balance and at a certain angle of incidence the component parallel to the wind direction is 102 N and the component normal to the wind direction is 4370 N. Determine the lift and drag coefficients of the aerofoil.
Answer $1\cdot24$, $0\cdot029$

12 An aircraft uses a certain aerofoil wing section of 3 m chord and 15 m total length. At an altitude of 5 km the Reynolds number for flow around the wing is 10^7 based on the chord. The aircraft weighs 120 kN, 90 per cent of which is supported by the wing. Determine the angle of incidence of the aerofoil and also the power required to propel the aircraft assuming that 20 per cent of the total drag comes from the wing.

Assume that the International Standard Atmosphere applies and that the following values of lift and drag coefficients can be used.

C_L	0·1	0·4	0·8	1·25	1·60	1·60
C_D	0·008	0·005	0·008	0·015	0·024	0·034
Angle of incidence (degrees)	−5	0	+5	+10	+15	+20

Answer 9·5°, 475 kW

10

Waterhammer and pressure transients

When the fluid flowing in a pipeline is brought to rest, for example by closing a valve, its momentum is destroyed and there will be a rapid rise of pressure of an amount depending on the velocity of flow, the retardation and the physical characteristics of the fluid and the pipeline. The following examples show different methods used to calculate the very large pressure rise which can occur.

10.1 Momentum theory for uniform retardation

> The outlet valve of a pipe 1200 m long is closed in 1 s causing a uniform retardation of the water flowing in the pipeline. If the initial velocity of the water is 1·8 m/s calculate the rise in pressure at the valve, assuming that the pipe is rigid and the fluid incompressible, from consideration of the loss of momentum.

Solution.

$$\text{Initial velocity} = v = 1\cdot8\,\text{m/s}$$
$$\text{Time of closure} = t = 1\,\text{s}$$
$$\text{Uniform retardation of water} = f = \frac{v}{t} = 1\cdot8\,\text{m/s}^2$$
$$\text{Mass of water in pipe} = \rho A L$$

where A = area of pipe, L = length of pipe.

$$\text{Force on valve} = \text{mass of water} \times \text{retardation}$$
$$= \rho A L f$$
$$\text{Pressure at valve} = \frac{\text{force}}{\text{area}} = \rho L f$$

Putting $\rho = 1000\,\text{kg/m}^3$ and $L = 1200\,\text{m}$,

$$\text{Pressure at valve} = 1000 \times 1200 \times 1\cdot8\,\text{N}$$
$$= 2160000\,\text{N} = \mathbf{2160\,kN}$$

10.2 Momentum theory for uniform rate of reduction of gate area

A reservoir discharges through a pipeline 900 m long and 3 m in diameter. The flow is controlled by a gate at the lower end of the pipe which is 150 m below the surface of the water in the reservoir. The initial flow is 42 m³/s which is reduced to 14 m³/s by reducing the gate area uniformly in 12 s. Calculate the maximum head rise at the gate and at the mid-point of the pipeline. Neglect friction loss in the pipe and assume a constant coefficient of discharge for the gate.

Solution In Fig. 10.1 let

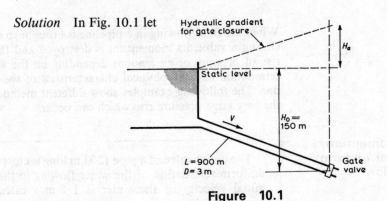

Figure 10.1

Solution. In Fig. 10.1 let

$$H_a = \text{head rise at gate due to closure}$$
$$H_0 = \text{static head}$$
$$A = \text{area of pipe}$$
$$V = \text{velocity in the pipe}$$

By Newton's second law

$$\rho g H_a \times A = -\rho A L \frac{dV}{dt}$$

$$\text{Head rise} = H_a = -\frac{L}{g}\frac{dV}{dt} \tag{1}$$

Before closure commences:

$$\text{Discharge to atmosphere} = Q_0 = AV_0 = (C_d A_g)_0 \sqrt{(2gH_0)}$$

where
$$A_g = \text{area of gate}$$
$$C_d = \text{gate discharge coefficient}$$

or
$$V_0 = B_0\sqrt{H_0}$$

where
$$B_0 = (C_d A_g)_0\sqrt{(2g)}/A.$$

At any instant during closure of gate:

$$\text{Instantaneous velocity in pipe} = V = B\sqrt{(H_0 + H_a)}$$

where $B = (C_d A_g)\sqrt{(2g)}/A$ at the given instant.

Therefore

$$\frac{V}{V_0} = \frac{B}{B_0} \sqrt{\left(1 + \frac{H_a}{H_0}\right)}$$

or if $B/B_0 = \tau$

$$V = \tau V_0 \sqrt{\left(1 + \frac{H_a}{H_0}\right)} \tag{2}$$

For uniform partial or complete gate closure.

Area after time $t = A_{gt} = A_{g0} - \delta A \frac{t}{T}$

where $\delta A =$ total area reduction in time T for completion of the change.
Therefore

$$\tau = \frac{B}{B_0} = \frac{A_{gt}}{A_{g0}} = 1 - \frac{\delta A}{A_{g0}} \frac{t}{T}$$

if C_d is constant.

If area of gate is equal to area of pipe when fully open

$$A_{g0} = A \quad \text{and} \quad \tau = 1 - \frac{\delta A}{A} \frac{t}{T}$$

Original discharge $= A V_0$

If $V_T =$ final velocity in pipe at completion of change

Final discharge $= A V_T$

Reduction of discharge $= A V_0 - A V_T$

$\qquad\qquad\qquad = $ loss of discharge through δA under steady
$\qquad\qquad\qquad\qquad$ conditions

$\qquad\qquad\qquad = \delta A V_0$

$\therefore \quad A(V_0 - V_T) = \delta A V_0$

$$\frac{\delta A}{A} = \frac{V_0 - V_T}{V_0}$$

Or putting $V_0 - V_T = V'$, $\dfrac{\delta A}{A} = \dfrac{V'}{V_0}$

Thus

$$\tau = 1 - \frac{V't}{V_0 T}$$

Substituting in equation (2)

$$V = \left(V_0 - \frac{V't}{T}\right) \sqrt{\left(1 + \frac{H_a}{H_0}\right)} \tag{3}$$

Solving equations (1) and (3) simultaneously for H_a and then putting $dH_a/dt = 0$ to find $(H_a)_{max}$ the maximum head rise at the gate

$$\frac{(H_a)_{max}}{H_0} = \frac{K}{2} + \sqrt{\left(K + \frac{K^2}{4}\right)}$$

where
$$K = \left(\frac{LV'}{gH_0T}\right)^2$$

Note. For uniform rate of gate *opening* $(H_a')_{\text{max}}$, the maximum head drop at the gate is given by

$$\frac{(H_a')_{\text{max}}}{H_0} = \frac{K}{2} - \sqrt{\left(K + \frac{K^2}{4}\right)}$$

In the present problem

Initial velocity of flow $= V_0 = \dfrac{Q_0}{A} = \dfrac{42}{\frac{1}{4}\pi \times 3^2} = 5\cdot945\,\text{m/s}$

Final velocity of flow $= V_T = \dfrac{Q_T}{A} = \dfrac{14}{\frac{1}{4}\pi \times 3^2} = 1\cdot982\,\text{m/s}$

$$V' = V_0 - V_T = 5\cdot945 - 1\cdot982 = 3\cdot963\,\text{m/s}$$
$$L = 900\,\text{m}, \quad H_0 = 150\,\text{m}, \quad T = 12\,\text{s}$$

Therefore

$$K = \left(\frac{LV'}{gH_0T}\right)^2 = \left(\frac{900 \times 3\cdot963}{9\cdot81 \times 150 \times 12}\right)^2 = 0\cdot0408$$

From equation (4)

$$\frac{(H_a)_{\text{max}}}{H_0} = \frac{0\cdot0408}{2} + \sqrt{\left(0\cdot0408 + \frac{0\cdot0017}{4}\right)}$$

$$= 0\cdot0204 + \sqrt{0\cdot0412} = 0\cdot224$$

Max head rise at the gate $= (H_a)_{\text{max}}$
$$= 0\cdot224 \times 150 = \mathbf{33\cdot56\,m}$$

Since H_a is proportional to L, the hydraulic gradient is a straight line and therefore the maximum head rise at the mid-point will be

$$\tfrac{1}{2}(H_a)_{\text{max}} = \mathbf{16\cdot78\,m}$$

10.3 Sudden closure in a rigid pipe

Describe the effect of the sudden closure of a valve at the end of an hydraulic pipeline.

If the initial velocity in the pipeline is $1\cdot5$ m/s, determine the rise in pressure when the valve is suddenly closed. Neglect the elasticity of the pipe walls but take into account the compressibility of the water in the pipe for which the bulk modulus is $2\cdot14 \times 10^9\,\text{N/m}^2$.

Solution. In the case of closure which is almost instantaneous the pressure generated at the valve is so great that the compressibility of the fluid must be taken into account and the method of example **10.1** is no longer suitable. When the valve is closed instantaneously the fluid immediately in contact with the valve is brought to rest and its kinetic energy is converted to volumetric strain energy of compression of the fluid. For a unit mass, if $v =$ initial velocity, $K =$ bulk modulus, $\rho =$ mass density and rise in pressure $= \delta p$, then

$$\text{Change in volume} = \frac{\delta p}{K} \times \frac{1}{\rho}$$

$$\text{Mean rise in pressure} = \tfrac{1}{2}\delta p$$

$$\text{Strain energy of compression} = \frac{\text{change in volume} \times \text{mean}}{\text{rise in pressure}}$$

$$= \frac{(\delta p)^2}{2K\rho}$$

Original kinetic energy of fluid per unit mass $= v^2/2$

so that
$$\frac{v^2}{2} = \frac{(\delta p)^2}{2K\rho}$$

or
$$\delta p = v\sqrt{(K\rho)}$$

The sequence of events for a sudden closure is:

(1) The fluid immediately in contact with the valve is brought to rest causing a pressure rise δp. A pressure wave travels back along the length L of the pipe as successive particles come to rest and it reaches the inlet end at time L/a where $a =$ velocity of sound in the fluid. The whole contents of the pipe is now stationary and at a pressure δp above normal.

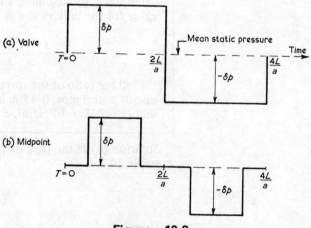

Figure 10.2

Figure 10.3

(2) As there is no resistance at the inlet end the increased pressure in the pipe δp will cause the particles at the inlet end to move outwards thus relieving the excess pressure. This effect travels back along the pipe reaching the closed valve at time $2L/a$ when the contents of the pipe will be at normal pressure and moving away from the valve with a velocity equal and opposite to the original velocity.

(3) The fluid tends to leave the valve but will adhere to it if the pressure does not fall to a level at which air or vapour is released. The particles of fluid are therefore successively brought to rest and there is a fall of pressure δp below normal which travels back as a pressure wave reaching the inlet end at time $3L/a$ when the whole column will be stationary and at reduced pressure.

(4) The fluid now surges back into the pipe again and a wave of normal pressure travels back down the pipe reaching the valve at time $4L/a$ when the whole column is moving towards the valve in its original state. The cycle of events will now repeat.

Figure 10.2 shows the conditions in the pipe at various times during the cycle and Fig. 10.3 shows the variation of pressure with time (a) at the valve and (b) at the mid-point of the pipeline.

When $v = 1\cdot 5\,\text{m/s}$, $K = 2\cdot 14 \times 10^9\,\text{N/m}^2$ and $\rho = 1000\,\text{kg/m}^3$

$$\text{Pressure rise } \delta p = v\sqrt{(K\rho)} = 1\cdot 5 \times \sqrt{(2\cdot 14 \times 10^{12})}\,\text{N/m}^2$$
$$= 2\cdot 194 \times 10^6\,\text{N/m}^2$$
$$= 2194\,\text{kN/m}^2$$

10.4 Elasticity of pipe

A liquid of mass density ρ and bulk modulus K flows with a uniform velocity v along a thin-walled pipe of internal diameter d and wall thickness t. E and σ denote the modulus of elasticity and Poisson's ratio for the material of which the pipe is made. Using these symbols, derive expressions for the sudden rise in pressure due to rapid closing of a valve (a) assuming the pipe to be non-elastic, (b) allowing for the elasticity of the pipe.

Show that for a given pipe the rise in pressure can be determined by assuming the pipe rigid, but using a modified value for the bulk modulus K_1, where

$$\frac{1}{K_1} = \frac{1}{K} + \frac{d}{4Et}(5 - 4\sigma)$$

Find the ratio of the correct rise to the approximate rise in the case of a steel pipe, $0\cdot 15$ m internal diam and $0\cdot 01$ m thick given that $E = 205 \times 10^9\,\text{N/m}^2$, $\sigma = 0\cdot 29$, and $K = 2\cdot 05 \times 10^9\,\text{N/m}^2$.

Solution (a) If the pipe is rigid the pressure rise is calculated as in example **10.3** and

$$\delta p = v\sqrt{K\rho} \tag{1}$$

(b) If the pipe is elastic the force on the valve due to sudden closure will produce a longitudinal stress f_L and a circumferential stress f_C in the pipe wall. The original kinetic energy of the fluid will be converted partly into strain energy of the fluid and partly into strain energy of the pipe. Considering a length of pipe x,

Volume of fluid in length $x = \frac{1}{4}\pi d^2 x$

Original kinetic energy of fluid $= \frac{1}{4}\pi d^2 x \cdot \rho \frac{1}{2} v^2$

Strain energy of fluid $= \frac{1}{2}\delta p \times$ change in volume

$$= \frac{1}{2}\delta p \times \frac{\delta p}{K} \times \frac{1}{4}\pi d^2 x$$

Longitudinal strain of pipe $= \dfrac{f_L}{E} - \sigma \dfrac{f_C}{E}$

Circumferential strain of pipe $= \dfrac{f_C}{E} - \sigma \dfrac{f_L}{E}$

Strain energy per unit volume of pipe wall

$$= \Sigma \tfrac{1}{2} \text{ stress} \times \text{strain}$$

$$= \tfrac{1}{2} f_L \left(\frac{f_L}{E} - \sigma \frac{f_C}{E} \right) + \tfrac{1}{2} f_C \left(\frac{f_C}{E} - \sigma \frac{f_L}{E} \right)$$

$$= \frac{1}{2E}(f_L{}^2 + f_C{}^2 + 2f_C f_L)$$

For a thin-walled pipe $f_L = \dfrac{\delta p d}{4t}$ and $f_C = \dfrac{\delta p d}{2t}$

Strain energy per unit volume of pipe

$$= \frac{1}{2E} \left(\frac{(\delta p)^2 d^2}{16t^2} + \frac{(\delta p)^2 r^2}{4t^2} - \sigma \frac{(\delta p)^2 r^2}{4t^2} \right)$$

Volume of pipe wall in length $x = \pi d t x$

Strain energy of pipe of length $x = \dfrac{\pi d^3 (\delta p)^2 x}{16Et}(5 - 4\sigma)$

Original k.e. of fluid = strain energy of fluid + strain energy of pipe

$$\tfrac{1}{4}\pi d^2 x \, \rho \tfrac{1}{2} v^2 = \tfrac{1}{4}\pi d^2 x \frac{(\delta p)^2}{2K} + \tfrac{1}{4}\pi d^2 x \frac{(\delta p)^2 d}{4Et}(5 - 4\sigma)$$

$$\delta p = v \sqrt{\frac{\rho}{\left[\dfrac{1}{K} + \dfrac{d}{4tE}(5 - 4\sigma) \right]}} \qquad (2)$$

If the pipe were assumed to be rigid and the liquid to have a modified bulk modulus K_1 then from equation (1)

$$\delta p = v \sqrt{K_1 \rho}$$

Comparing this with equation (2)

$$\frac{1}{K_1} = \frac{1}{K} + \frac{d}{4tE}(5 - 4\sigma)$$

Ratio of correct rise to approximate rise

$$= \frac{v\sqrt{K_1\rho}}{v\sqrt{K\rho}}$$

$$= \sqrt{\frac{K_1}{K}} = \sqrt{\frac{1}{K \times \dfrac{1}{K_1}}}$$

$$= \frac{1}{\sqrt{\left[1 + \dfrac{Kd}{4tE}(5 - 4\sigma)\right]}}$$

$$= \frac{1}{\sqrt{\left[1 + \dfrac{2\cdot05 \times 10^9 \times 0\cdot15}{4 \times 0\cdot01 \times 205 \times 10^9}(5 - 4 \times 0\cdot29)\right]}}$$

$$= \frac{1}{\sqrt{(1 + 0\cdot144)}} = \textbf{1 to 1}\cdot\textbf{07}$$

10.5 Velocity of a pressure wave

Derive an expression for the velocity of transmission of a pressure wave through a fluid of bulk modulus K and mass density ρ.

What will be the velocity of sound through water if $K = 2\cdot05 \times 10^9$ N/m² and $\rho = 1000$ kg/m³?

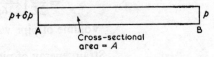

Cross-sectional
area = A

Figure 10.4

Solution. Consider a length of fluid (Fig. 10.4) of cross-sectional area A in which the pressure at A is $p + \delta p$ and at B is p. Let a be the
In unit time, distance travelled by wave = a
∴ Volume of fluid compressed in unit time = Aa

$$\text{Change of volume of fluid} = Aa\,\frac{\delta p}{K}$$

$$\text{Velocity of A towards B} = a\,\frac{\delta p}{K}$$

$$\text{Mass compressed in unit time} = Aa\rho$$

Applying Newton's second law

$$A\delta p = Aa\rho \times a\,\frac{\delta p}{K}$$

$$a = \sqrt{\frac{K}{\rho}}$$

Since sound is a pressure wave, a is the velocity of sound.

Putting $\qquad K = 2\cdot 05 \times 10^9\,\text{N/m}^2$

and $\qquad \rho = 1000\,\text{kg/m}^3$

then $\qquad a = \sqrt{\dfrac{2\cdot 05 \times 10^9}{10^3}} = \mathbf{1432\,m/s}$

10.6 Slow closure without reflection

(a) Prove that the pressure increase produced in a pipe by sudden decrease of velocity dv due to closure of a valve is

$$dP = \rho a dv$$

where ρ is the density of water and a is the velocity of sound transmission through water in the pipe concerned.

(b) A valve at the outlet end of a pipe 1800 m long through which water flows at $2\cdot 5$ m/s is closed in 2 s in such a way that the retardation is uniform. Assuming that $a = 1372$ m/s, find the increase of pressure caused by the valve closure.

Solution. (a) The pressure wave due to sudden decrease of velocity δv will be propagated along the pipe with a velocity a. If $A =$ cross-sectional area of pipe and $\rho =$ mass density of fluid,

Distance travelled by pressure wave in one sec $= a$

$\therefore$ Mass undergoing change of velocity δv in one sec $= a$

Force due to pressure rise $= A\delta P$

$\qquad\qquad\qquad\qquad = $ rate of change of momentum

Thus $\qquad A\delta P = \rho\,Aa \times \delta v$

$$\delta P = \rho a \delta v$$

(b) Let $T =$ time of closure of valve,

$\qquad v =$ initial velocity of flow.

If T is less than L/a the valve will close before the pressure wave reaches the open end of the pipe.

Distance travelled by wave $= s = Ta$

If retardation is uniform, by Newton's second law

$$A\delta p = \rho As \times \frac{v}{T} \quad \text{where} \quad A = \text{pipe area}$$

$$= \rho As \times \frac{va}{s}$$

$$\delta p = \rho av = v\sqrt{(K\rho)}$$

This is the same pressure rise as for instantaneous closure. For values of T between L/a and $2L/a$ the pressure at the valve will also be substantially that for instantaneous closure since the wave of low pressure reflected from the open end does not reach the valve until $T = 2L/a$ (*see* example **10.3**).

Putting $L = 900$ m and $a = 1372$ m/s

$$\frac{L}{a} = \frac{1800}{1372} = 1\cdot31\,\text{s}$$

Since closure occurs in 2 s the reflected wave, which does not reach the valve until $2L/a = 2\cdot62$ s after commencement of closure, does not affect the result.

Putting $\rho = 1000\,\text{kg/m}^3$, $a = 1372$ m/s, $v = 2\cdot5$ m/s.

$$\delta p = 10^3 \times 1372 \times 2\cdot5\,\text{N/m}^2$$

Pressure rise $= \delta p = \mathbf{3430\,kN/m^2}$

10.7 Reflection of pressure wave at open end and at a dead end

Derive from first principles the fundamental waterhammer equations.

From these equations deduce (*a*) an expression for the rise of pressure at a gate at the outlet of a pipeline when it is closed instantaneously, (*b*) the changes which occur when a pressure wave is reflected (i) at an open reservoir, (ii) at a dead end.

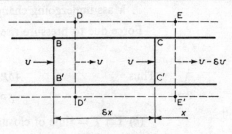

Figure   10.5

Solution. When a column of compressible fluid flowing in an elastic pipe is decelerated by closure of a valve at the outlet, the fluid immediately adjacent to the valve is brought to rest by an increase of pressure at the valve. Since the fluid is compressible and the pipe elastic the remainder of the column of fluid will not be affected immediately, but the rise of pressure will travel back along the pipe as each section of the fluid undergoes deceleration in succession.

By Newton's second law, considering an element of length δx (Fig. 10.5), if δh is the pressure rise due to deceleration from v to $v - \delta v$ in time δt in a pipe of cross-sectional area A,

Force due to pressure rise = rate of change of momentum

$$A\rho g\delta h = \rho A\delta x \frac{\delta v}{\delta t}$$

or

$$\frac{dh}{dx} = -\frac{1}{g}\frac{dv}{dt} \tag{1}$$

The fluid compresses and the pipe expands, but the fluid must still fill the available space. An element occupying the volume BB'C'C at time t will as it is decelerated occupy the volume DD'E'E at time $t + \delta t$ and the velocity at DD' will be v and at EE' will be $v - \delta v$

Reduction in the length of the element

$$= v\delta t - (v - \delta v)\delta t = \delta v\delta t$$

Change in volume $= -A\delta v\delta t$

Volumetric strain $= -\dfrac{A\delta v\delta t}{A\delta x} = -\dfrac{\delta v}{\delta x}\delta t$

Pressure stress $= \rho g\delta h$

If K_1 is the apparent bulk modulus of the fluid as modified to allow for the elasticity of the pipe (*see* example **10.4**),

$$\text{Volumetric strain} = \frac{\text{pressure stress}}{K_1}$$

$$-\frac{\delta v}{\delta x}\delta t = \frac{\rho g\delta h}{K_1}$$

and

$$\frac{\delta h}{\delta t} = -\frac{\delta v}{\delta x}\frac{K_1}{\rho g}$$

but $K_1/\rho = a^2$ where a is the velocity of propagation of the pressure wave. Therefore

$$\frac{dh}{dt} = -\frac{a^2}{g}\frac{dv}{dx} \tag{2}$$

The general solution of equations (1) and (2) when x is measured from the downstream end of the pipe (in the opposite direction to v) are:

$$H - H_0 = F\left(t - \frac{x}{a}\right) + f\left(t + \frac{x}{a}\right) \tag{3}$$

$$v - v_0 = -\frac{g}{a}\left\{F\left(t - \frac{x}{a}\right) - f\left(t + \frac{x}{a}\right)\right\} \tag{4}$$

where $F(t - x/a)$ and $f(t + x/a)$ are the mean functions of $(t - x/a)$ and $(t + x/a)$ respectively and H_0 and v_0 are the values of H and v when $t = 0$.

From the dimensions of equation (3), $F(t - x/a)$ and $f(t + x/a)$

must have the dimensions of pressure heads and represent pressure waves.

Abbreviating $F(t - x/a)$ to F and $f(t + x/a)$ to f, the F pressure wave moves in the direction of positive x with velocity a, while the f wave moves in the direction of negative x with velocity a. Since F is a function of $(t - x/a)$ and f is a function of $(t + x/a)$ the two waves are out of phase by $2x/a$, the f wave lagging behind F. When the two waves pass neither is altered.

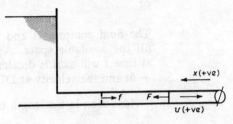

Figure 10.6

The pressure rise at a given point x on the pipe line at a given time t is the algebraic sum of the pressure head of the F and f waves.

Equations (3) and (4) are the fundamental waterhammer equations.

(a) If the gate is closed instantaneously there is no reflected wave f at the instant of closure and equations (3) and (4) become

$$H - H_0 = F\left(t - \frac{x}{a}\right)$$

$$v - v_0 = -\frac{g}{a}F\left(t - \frac{x}{a}\right)$$

Eliminating $F\left(t - \frac{x}{a}\right)$

$$H - H_0 = -\frac{a}{g}(v - v_0)$$

$$\text{Head rise} = \delta H = -\frac{a}{g}\delta v$$

or pressure rise $= \delta p = -\rho a \delta v$ (*see* example **10.5**).

(b) (i) *Reflection at open reservoir.* A movement of the control gate causes an F wave to originate at the gate and move with velocity a in the x direction reaching the reservoir at time L/a where L = length of the pipe. On reaching the reservoir a reflected wave of the f type is produced.

The pressure head at the open reservoir is not altered by the transient pressure in the pipe and, equation (3) is therefore

$$H - H_0 = 0 = F\left(t - \frac{L}{a}\right) + f\left(t + \frac{L}{a}\right)$$

$$\therefore \quad F\left(t - \frac{L}{a}\right) = -f\left(t + \frac{L}{a}\right) \tag{5}$$

substituting in equation (4)

$$v - v_0 = \frac{2g}{a} F\left(t - \frac{L}{a}\right) \tag{6}$$

Equation (5) indicates that when the F type wave reaches the reservoir it produces an f wave of equal magnitude but opposite sign. Equation (6) indicates that the change of velocity at the reservoir is twice that produced by the F wave at other points thus giving rise to the f wave.

The reflected wave reaches the gate at time $2L/a$ after the F wave which produced it and is equal in magnitude and opposite in sign to the F wave.

(ii) *Reflection at a dead-end.* At any time t, $v = v_0 = 0$ at the dead-end, where $x = L$.

Equation (4) now gives

$$F\left(t - \frac{L}{a}\right) = f\left(t + \frac{L}{a}\right)$$

substituting in equation (3)

$$H - H_0 = 2F\left(t - \frac{L}{a}\right) \tag{7}$$

Equation (7) indicates that the pressure wave is entirely reflected without change of sign and the pressure rise equals twice the intensity of the direct pressure wave.

10.8 Reflection at a partially open gate

Show from consideration of the fundamental waterhammer equations that when a pressure wave reaches a partially open gate it will be partially reflected.

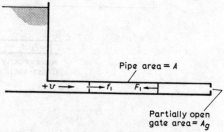

Figure 10.7

Solution. In Fig. 10.7 let f_1 be a pressure wave moving towards the partially open gate and F_1 the wave produced by reflection of the f_1 wave at the gate. Then

$$H - H_0 = F_1 + f_1 \tag{1}$$

and

$$v - v_0 = -\frac{g}{a}(F_1 - f_1) \tag{2}$$

Considering the discharge Q through the gate

$$Q = Av = C_d A_g \sqrt{(2gH)}$$

$$\therefore \quad v = B\sqrt{H} \tag{3}$$

where

$$B = \frac{C_d A_g}{A} \sqrt{(2g)}$$

Solving equations (1), (2) and (3)

$$v = -\frac{aB^2}{2g} + \frac{B}{2}\sqrt{\left[\left(\frac{aB}{2g}\right)^2 + 4\left(H_0 + \frac{av_0}{g} + 2f_1\right)\right]} \tag{4}$$

and from equation (2)

$$F_1 = -\frac{a}{g}(v - v_0) + f_1 \tag{5}$$

If equations (4) and (5) are applied to an open gate it is found that F_1 is always numerically less than f_1 indicating that the pressure waves are partially reflected at the gate.

10.9 Reflection and transmission at change of section

Two pipes BC and CD are of different cross-sectional areas A_1 and A_2 and also differ in thickness and material so that the velocities of propagation of a pressure wave in the two pipes are V_{p1} and V_{p2} respectively (*see* Fig. 10.8). A pressure wave originating from B will be partially reflected and partially transmitted at C, derive expressions for the transmission and reflection factors at C.

A pipe of 0·5 m diameter is connected in series with a pipe of 0·75 m diameter. The velocity of propagation is 1300 m/s in the small pipe and 1350 m/s in the large pipe. If a pressure wave due to a sudden change of head of 60 m travels towards the junction along the smaller pipe, what will be the magnitude of the reflected and transmitted waves at the junction?

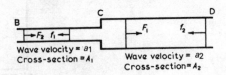

B   C   D

$\rightarrow F_2 \ \ f_1 \leftarrow$ $\rightarrow F_1$ $f_2 \leftarrow$

Wave velocity = a_1
Cross-section = A_1

Wave velocity = a_2
Cross-section = A_2

Figure   10.8

Solution. Let C′ indicate a point at C just inside BC and C″ indicate a point at C just inside CD.

Applying the waterhammer equations at time t for an F_1 wave reaching C:

At C′ $H_{c't} - H_{c'o} = F_1 + f_1$

$$v_{c't} - v_{c'o} = -\frac{g}{a_1}$$

At C″
$$H_{c''t} - H_{c'o} = F_2$$

$$v_{c''t} - v_{c'o} = -\frac{g}{a_2} F_2$$

For continuity of flow at C

$$A_2 v_{c''t} = A_1 v_{c't}$$

From these equations

$$F_2 = sF_1 \text{ where } s = \frac{2A_1/a_1}{A_1/a_1 + A_2/a_2} = \text{transmission factor}$$

$$f_1 = rF_1 \text{ where } r = \frac{A_1/a_1 - A_2/a_2}{A_1/a_1 + A_2/a_2} = \text{reflection factor}$$

and
$$s - r = 1$$

Thus the transmission and reflection factors depend on cross-sectional area and on the velocity of wave propagation. Since the wave velocity depends on the pipe thickness and elastic properties as well as on the properties of the fluid (see example **10.4**) a wave reflection will occur at changes of pipe thickness or material as well as at changes of section.

Putting
$$A_1 = \tfrac{1}{4}\pi(0.5)^2 = 0.1963\,\text{m}^2$$

$$A_2 = \tfrac{1}{4}\pi(0.75)^2 = 0.4416\,\text{m}^2$$

$$F_1 = 60\,\text{m}, \quad a_1 = 1300\,\text{m/s}, \quad a_2 = 1250\,\text{m/s}$$

$$\frac{A_1}{a_1} = \frac{0.1963}{1300} = 1.51 \times 10^{-4}$$

$$\frac{A_2}{a_2} = \frac{0.4416}{1350} = 3.27 \times 10^{-4}$$

$$s = \frac{3.02 \times 10^{-4}}{(1.51 + 3.27) \times 10^{-4}} = \frac{3.02}{4.78} = 0.632$$

$$\therefore \qquad F_2 = 0.632 \times 60 = \mathbf{37.9\,m}$$

$$r = \frac{1.51 - 3.27}{4.78} = \frac{-1.76}{4.78} = -0.368$$

$$\therefore \qquad f_1 = -0.368 \times 60 = \mathbf{-22.1\,m}$$

Flow in a pipeline is controlled by the closure of a gate at the lower end. Assuming that this gate is closed slowly in a time greater than $2L/a$, where L is the length of the pipeline and a the velocity of propagation of a pressure wave in the system, derive from first principles the equations for head rise and velocity of flow at any given time.

A pipeline has a length L of 1200 m and diameter D of 3 m. The flow in the line is controlled by a gate at the lower end. The initial velocity of flow v_0 is $3 \cdot 27$ m/s, the initial head H_0 at the gate is 150 m and the velocity of propagation of a pressure wave a is 1200 m/s. The gate is closed slowly in 6 s and the area of the gate opening A_g multiplied by the coefficient of discharge C_d varies as follows:

t sec	0	1	2	3	4	5	6
$C_d A_g \text{m}^2$	$0 \cdot 437$	$0 \cdot 393$	$0 \cdot 306$	$0 \cdot 218$	$0 \cdot 131$	$0 \cdot 044$	0

Plot a graph showing the variation of pressure at the valve with time over the period 0 to 12 s and determine the maximum pressure rise at the valve.

Solution. For slow closures, in which the time of closure is less than the time $2L/a$ required for a wave to travel from the valve to the open end of the pipe, be reflected and return to the valve, the actual reduction of valve area as it closes is usually considered as a series of instantaneous steps each comprising a sudden partial closure and generating a pressure wave. For convenience of calculation the interval between steps is made as large as possible When considering conditions at the gate an interval of $2L/a$ is usually taken. For other points on the pipeline the largest time interval that can be used is the shortest time for a wave to travel from the point under consideration to a point where it is reflected and then return.

From example **10.7** equations (3) and (4)

$$\text{Head rise at valve at time } t = H - H_0 = F + f \tag{1}$$

and

$$v - v_0 = \frac{-g}{a}(F - f) \tag{2}$$

where $v =$ velocity in the pipe at time t. Also as shown in example **10.8** for partial closure

$$v = -\frac{aB^2}{2g} + \frac{B}{2}\sqrt{\left[\left(\frac{aB}{g}\right)^2 + 4\left(H_0 + \frac{av_0}{g} + 2f\right)\right]} \tag{3}$$

where

$$B = \frac{C_d A_g}{A}\sqrt{(2g)} \tag{4}$$

and the value of the reflected wave from the open end for the nth step will be

$$f_n = -F\left(n - \frac{2L}{a}\right) \tag{5}$$

From equation (2),

$$F = -\frac{a}{g}(v - v_0) + f \tag{6}$$

From equations (1) to (6) the values of F, f and the head rise $(F + f)$ can be found for each interval of time. The work can be set out in tabular form. Although

$$\frac{2L}{a} = \frac{2 \times 1200}{1200} = 2\,\text{s}$$

the information for gate closure is given for 1s intervals and Table 10.1 is prepared on this basis.

Table 10.1

(1) t secs	(2) $C_d A_g$ metres	(3) B	(4) v m/s	(5) F metres	(6) f metres	(7) Headrise $(F + f)$ metres
0	0·437	0·2740	3·270	0	0	0
1	0·393	0·2464	3·156	+13·99	0	+13·99
2	0·306	0·1919	2·779	+60·01	0	+60·01
3	0·218	0·1367	2·185	+118·74	−13·99	+104·75
4	0·131	0·0821	1·337	+176·46	−60·01	+116·45
5	0·044	0·0276	0·4435	+227·00	−118·74	+108·26
6	0	0	0	+223·54	−176·46	+47·08
7	0	0	0	+173·00	−227·00	−54·00
8	0	0	0	+176·46	−223·54	−47·08
9	0	0	0	+227·00	−173·00	+54·00
10	0	0	0	+223·54	−176·46	+47·08
11	0	0	0	+173·00	−227·00	−54·00
12	0	0	0	+176·46	−223·54	−47·08

Columns 1 and 2 contain the original data for closure of the gate.
Column 3: values of B are calculated from equation (4)

$$B = \frac{C_d A_g}{A}\sqrt{(2g)} = (C_d A_g)\frac{\sqrt{(19\cdot62)}}{\frac{1}{4}\pi \times 3^2} = 0\cdot627(C_d A_g)$$

Column 4: values of v are calculated from eqn. (3):

$$v = -\frac{1200}{19\cdot62}B^2 + \frac{B}{2}\sqrt{\left[\left(\frac{1200}{9\cdot81}\right)^2 B^2 + 4\left(150 + \frac{1200 \times 3\cdot27}{9\cdot81} + 2f\right)\right]}$$

$$= -61\cdot16B^2 + \frac{B}{2}\sqrt{[14963B^2 + 2200 + 8f]}$$

Column 5: Values of F are calculated from equation (6):

$$F = -\frac{1200}{9 \cdot 81}(v - 3 \cdot 27) + f$$

$$= -122 \cdot 32(v - 3 \cdot 27) + f$$

Column 6: values of f are obtained from equation (5). The value of f is that for F from the step $2L/a$ previous but of opposite sign as shown by the arrow.

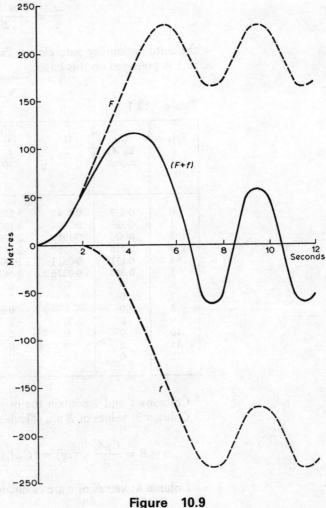

Figure   10.9

Figure 10.9 shows the values of F, f and the head rise $(F + f)$ plotted against time. Maximum head rise at the gate = **116·45m.**

(*a*) Derive the conjugate waterhammer equations and (*b*) show their application to the graphical determination of the maximum head rise in the pipeline of Example **10.10**.

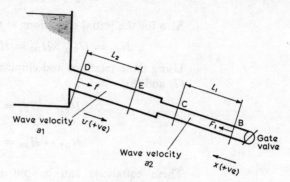

Figure 10.10

Solution (Fig. 10.10).
(*a*) From Example 10.7 equations (3) and (4) the basic equations are

$$H - H_0 = F\left(t - \frac{x}{a}\right) + f\left(t + \frac{x}{a}\right) \tag{1}$$

and

$$v - v_0 = -\frac{g}{a}\left\{F\left(t - \frac{x}{a}\right) - f\left(t + \frac{x}{a}\right)\right\} \tag{2}$$

Subtracting equation (2) from (1)

$$H - H_0 = \frac{a}{g}(v - v_0) + 2F\left(t - \frac{x}{a}\right) \tag{3}$$

Adding equations (1) and (2)

$$H - H_0 = -\frac{a}{g}(v - v_0) + 2f\left(t + \frac{x}{a}\right) \tag{4}$$

Referring to Fig. 10.10, the pipeline shown has a gate value at its outlet and the material, diameter and thickness of the two sections are such that the velocity of wave propagation is a_1 and a_2 respectively.

If there is an F wave at B at time t_1 it will move up the pipe with velocity a_1 and reach any point C at time $t_2 = t_1 + L_1/a_1$. Similarly an f wave situated at D at a time t_3 will move down the pipe with velocity a_2 and reach any point E at time $t_4 = t_3 + L_2/a_2$.

Applying equation (3) to B and C,

$$H_{B1} - H_{B0} = \frac{a_1}{g}(v_{B1} - v_{B0}) + 2F \tag{5}$$

$$H_{C1} - H_{C0} = \frac{a_1}{g}(v_{C1} - v_{C0}) + 2F \tag{6}$$

where H_{B1} indicates the value of H at B at time 1 at which the velocity is v_{B1}, the corresponding values at time 0 being H_{B0} and v_{B0}.

Using the same notation and applying equation (4) to D and E,

$$H_{D3} - H_{D0} = -\frac{a_2}{g}(v_{D3} - v_{D0}) + 2f \tag{7}$$

$$H_{E4} - H_{E0} = -\frac{a_2}{g}(v_{E4} - v_{E0}) + 2f \tag{8}$$

Also for the initial conditions at time 0,

$$H_{B0} = H_{C0}, \quad H_{D0} = H_{E0}, \quad v_{B0} = v_{C0}, \quad v_{D0} = v_{E0}$$

Using these relations and eliminating F and f from equations (5), (6), (7) and (8),

$$H_{B1} - H_{C2} = \frac{a_1}{g}(v_{B1} - v_{C2}) \tag{9}$$

$$H_{D3} - H_{E4} = -\frac{a_2}{g}(v_{D3} - v_{E4}) \tag{10}$$

These equations can be put in dimensionless form. Considering equation (9),

$$\frac{H_{B1} - H_{C2}}{H_0} = \frac{a_1}{g}\frac{v_0}{H_0}\left(\frac{v_{B1}}{v_0} - \frac{v_{C2}}{v_0}\right)$$

Putting $\quad \dfrac{H}{H_0} = \bar{h}, \quad \dfrac{v}{v_0} = \bar{v} \quad$ and $\quad \dfrac{a_1 v_0}{2gH_0} = \rho,$

$$\bar{h}_{B1} - \bar{h}_{C2} = 2\rho(\bar{v}_{B1} - \bar{v}_{C2}) \tag{11}$$

Similarly from equation (10),

$$\bar{h}_{D3} - \bar{h}_{E4} = -2\rho(\bar{v}_{D3} - \bar{v}_{E4}) \tag{12}$$

Equations (11) and (12) are the conjugate waterhammer equations. Equation (11) is for the F wave travelling upstream and equation (12) is for the f wave travelling downstream. On a plot of head ratio $\bar{h}$ against velocity ratio $\bar{v}$ equations (11) and (12) are straight lines with equal but opposite slopes 2ρ.

(b) In the problem of example **10.10** details of which are given in Fig. 10.11 the pipe is uniform so that in equations (11) and (12) points B and D can conveniently be taken at the valve and open end respectively. Point C can be taken to coincide with D and point E to coincide with B. The time interval between steps is

$$t = \frac{L}{a} = \frac{1200}{1200} = 1\,\text{s}$$

Equations (11) and (12) become

$$\bar{h}_{B1} - \bar{h}_{D2} = 2\rho(\bar{v}_{B1} - \bar{v}_{D2}) \tag{13}$$

$$\bar{h}_{D3} - \bar{h}_{B4} = -2\rho(\bar{v}_{D3} - \bar{v}_{B4}) \tag{14}$$

A series of similar equations can be written for each time interval. Each of these equations defines the relation between $\bar{h}$ and $\bar{v}$ at one end of the pipe in terms of the values of $\bar{h}$ and $\bar{v}$ at the other end at the next time interval. The additional information for the solution comes from the end conditions.

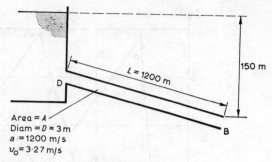

Figure 10.11

At D, the open end, the head remains constant and $\hbar_D = 1$.

At B the relation between v and H for discharge through the valve is known.

Initially
$$v_{B0} = B_0\sqrt{H_{B0}}$$

where $B_0 = \dfrac{C_d A_{g0}}{A}\sqrt{(2g)}$

A_{g0} = gate area at time 0

A = pipe area

C_d = discharge coefficient

For any other gate opening

$$v_B = B\sqrt{H_B}$$

from which
$$\bar{v}_B = \tau\sqrt{\hbar_B} \tag{15}$$

where $\tau = B/B_0$ *(see example 10.2).*

In the present case,

$$\text{Pipe area } A = \tfrac{1}{4}\pi \times 3^2 = 7 \cdot 06\,\text{m}^2$$

For the gate the conditions at each time interval are

Time t sec	0	1	2	3	4	5	6
$C_d A_g$ (m²)	0·437	0·393	0·306	0·218	0·131	0·044	0
B	0·2740	0·2464	0·1919	0·1367	0·0821	0·0276	0
$\tau = B/B_0$	1·000	0·900	0·700	0·498	0·300	0·101	0

The graphical solution is obtained using a plot of $\hbar$ against $\bar{v}$ for equation (13), (14) and (15) as follows (Fig. 10.12):

(i) Using a time interval of $t = L/a = 1$ s and taking the values of τ calculated above draw the parabolas representing equation (15) for $t = 0, 1, 2, 3, 4, 5$ and 6 s. The latter coincides with the $\hbar$ axis.

(ii) The head at the open end will always be constant and is represented for each value of t by points D_0, D_1, etc. which must lie on the line $\hbar = 1$.

(iii) The head at the valve $\hbar_B$ at each time t must lie on the corresponding parabola for that value of t. Thus at time $t = 1$ the

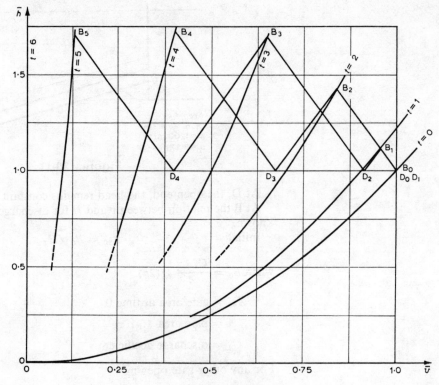

Figure 10.12

head $\bar{h}_{B1}$ must be at some point B_1 on the parabola for $t = 1$.

(iv) Commencing with the steady state conditions, the point B_0, before movement commences is plotted at $\bar{h} = 1$, $\bar{v} = 1$. At the open end D no change will occur until after $t = 1$ when the wave arrives at D, therefore D_0 and D_1 coincide with B_0.

(v) The head ratio $\bar{h}_{B1}$ at the gate at time $t = 1$ is found by starting from D_0 and finding the point B_1 where the parabola for $t = 1$ is intersected by the f line through D_0 of slope

$$-2\rho = -\frac{av_0}{gH_0} = -\frac{1200 \times 3 \cdot 27}{9 \cdot 81 \times 150} = -2 \cdot 67$$

(vi) Similarly B_2 is the intersection of the f line from D_1 with the parabola for $t = 2$.

(vii) B_3 will be at the intersection of the f line from D_2 with the parabola for $t = 3$ and D_2 is located by the intersection of the F line, slope 2ρ, drawn from B_1 to intersect $\bar{h} = 1$.

(viii) Similarly B_4 is at the intersection of the f line from D_3 which is located from B_2, and B_5 is found from D_4 which is located from B_3.

From the diagram (Fig. 10.12) it can be seen that h has a maximum value at B_4 where

$$h_{max} = \frac{H_{max}}{H_0} = 1 \cdot 74$$

Head rise at valve $= H_{max} - H_0 = 0 \cdot 74 H_0$

$$= 0 \cdot 74 \times 150 = \textbf{112m}$$

10.12 Surge tank

In a hydro-electric scheme the supply pipeline is $1 \cdot 2$ m diameter and has a resistance coefficient $f = 0 \cdot 01$. At 150 m along the pipeline from the reservoir there is an open surge tank $3 \cdot 6$ m diameter with no restriction. The steady full flow to the turbines is $2 \cdot 27$ m³/s.

(*a*) Develop the basic equations for mass oscillation in the surge tank due to a sudden change of flow.

(*b*) Show by giving a few stages of step-by-step integration how to estimate the maximum rise of the water level in the surge tank for a sudden full flow rejection.

Solution. Sudden changes in the demand of the turbines could cause high inertia pressures in the supply line. The function of a surge tank is to prevent excessive pressures occurring when the demand is reduced by providing a storage volume into which the flow can pass. As the level in the surge tank rises the back pressure increases and the flow in the pipeline is gradually decelerated. The surge tank also provides a subsidiary supply to ensure that sufficient water is immediately available close to the turbine when there is a sudden increase of load.

(*a*) The arrangement is shown in Fig. 10.13. When a sudden decrease

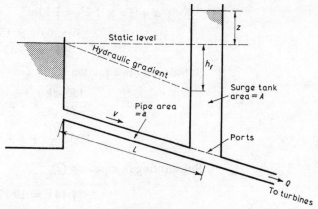

Figure   10.13

of flow occurs the water column in the pipe between the reservoir and the surge tank is retarded and the pressure at the base of the surge tank increases causing flow from the pipe into the tank. If at any time t the surface of the water in the surge tank is z above the static level and the friction head is h_f,

$$\text{Head opposing flow} = z \pm h_f$$

where h_f is positive when flow is into the tank and changes to negative when the flow reverses after the level in the tank has reached its peak.

$$\text{Force opposing motion of water in pipe} = \rho g a(z \pm h_f)$$

$$\text{Rate of change of momentum in pipe} = \rho a L \frac{dV}{dt}$$

If we assume that the water is incompressible and the pipe inelastic

$$\rho g a(z \pm h_f) = -\rho a L \frac{dV}{dt}$$

$$\frac{L}{g}\frac{dV}{dt} + z \pm h_f = 0 \tag{1}$$

where

$$h_f = \frac{4fL}{d}\frac{V^2}{2g} = CV^2$$

$$\text{Surface velocity of surge tank} = \frac{dz}{dt}$$

$$\therefore \text{Flow into surge tank} = A\frac{dz}{dt}$$

If V and Q are the pipe velocity and turbine demand at time t, for continuity of flow,

$$\text{Flow down pipe} = \text{flow into surge tank} + \text{flow to turbines}$$

$$aV = A\frac{dz}{dt} + Q \tag{2}$$

For the present problem,

$$a = \tfrac{1}{4}\pi \times 1\cdot2^2 = 1\cdot13\,\text{m}^2, \quad A = \tfrac{1}{4}\pi \times 3\cdot6^2 = 10\cdot17\,\text{m}^2$$

$$h_f = \frac{4 \times 0\cdot01 \times 150}{1\cdot2 \times 19\cdot62}V^2 = 0\cdot255\,V^2$$

Substituting in equation (1)

$$\frac{150}{9\cdot81}\frac{dV}{dt} + z \pm 0\cdot255\,V^2 = 0$$

$$15\cdot33\frac{dV}{dt} + z \pm 0\cdot255\,V^2 = 0 \tag{3}$$

Substituting in equation (2)

$$1\cdot13V = 10\cdot17\frac{dz}{dt} + Q$$

$$V = 9\frac{dz}{dt} + 0\cdot886Q \tag{4}$$

Equations (3) and (4) are the basic equations for this problem.

(b) Without resorting to a step-by-step solution an approximate estimate of the maximum surge level and the time at which it occurs could be obtained by neglecting friction.

Equation (1) becomes

$$\frac{L}{g}\frac{dV}{dt} + z = 0 \tag{5}$$

For total rejection $Q = 0$ and so from equation (2)

$$A\frac{dz}{dt} = aV \tag{6}$$

When $t = 0$, $z = 0$ and $\frac{dz}{dt} = Q_0/A$

where Q_0 = steady flow in pipe before change.
Solving equations (5) and (6) simultaneously

$$z = \frac{Q_0}{A}\sqrt{\left(\frac{A}{a}\frac{L}{g}\right)}\sin\sqrt{\left(\frac{a}{A}\frac{g}{L}\right)}\,t$$

giving

$$z_{\text{max}} = \frac{Q_0}{A}\sqrt{\left(\frac{A}{a}\frac{L}{g}\right)}$$

Putting $Q_0 = 2\cdot27\,\text{m}^3/\text{s}$, $A = 10\cdot17\,\text{m}^2$, $a = 1\cdot13\,\text{m}^2$, $L = 150\,\text{m}$,

$$z_{\text{max}} = \frac{2\cdot27}{10\cdot17}\sqrt{\left(\frac{10\cdot17}{1\cdot13}\times\frac{150}{9\cdot81}\right)} = 2\cdot62\,\text{m}$$

Periodic time of surge $= T = 2\pi\sqrt{\left(\frac{A}{a}\frac{L}{g}\right)}$

$$= 2\pi\sqrt{\frac{9\times150}{9\cdot81}} = 73\cdot8\,\text{s}$$

The surge follows a sine wave and so the maximum rise in level occurs at $\frac{1}{4}T = 18\cdot45\,\text{s}$ after the start.

A more accurate computation can be made using step-by-step integration of equations (3) and (4). For total rejection $Q = 0$ and replacing dz, dt and dV by finite intervals Δz, Δt and ΔV, equation (4) becomes

$$V = 9\frac{\Delta z}{\Delta t} \tag{7}$$

and equation (3), for upsurge, is

$$z = -\left(0\cdot255V^2 + 15\cdot3\frac{\Delta V}{\Delta t}\right) \tag{8}$$

At the start, when $t = 0$

$$V_0 = \frac{Q_0}{a} = \frac{2\cdot27}{1\cdot13} = 2\cdot01\,\text{m/s}$$

$$Z_0 = -0\cdot255V^2 = -0\cdot255\times2\cdot01^2 = -1\cdot03\,\text{m}$$

Working in steps of $\Delta t = 5\,\text{s}$, the mean values of V and z are worked out for each step. If V_i and z_i are the values of V and z at the beginning of each step, the mean values for the step are

$$V_m = V_i + \tfrac{1}{2}\Delta V \tag{9}$$

$$Z_m = z_i + \tfrac{1}{2}\Delta z \tag{10}$$

The procedure for each step is as follows

(i) Estimate Δz, then calculate V_m from equation (7)

$$V_m = 9\,\frac{\Delta z}{\Delta t}$$

(ii) Using this value of V_m calculate

$$\Delta V = 2(V_m - V_i)$$

Then from equation (8) calculate

$$z_m = -\left(0\!\cdot\!255 V_m{}^2 + 15\!\cdot\!3\,\frac{\Delta V}{\Delta t}\right)$$

(iii) Compare the value of z_m from step (ii) with the value of z_m from equation (10). If the two values do not agree re-estimate Δz and repeat the step.

If agreement is satisfactory proceed to the next step with initial conditions $z_i + \Delta z$ and $V_i + \Delta V$ from the previous step. The working is carried out as shown in Table 10.2. As the peak value of z is approached the time interval Δt is reduced.

Table 10.2

t secs	Δt	z_i metres	V_i m/s	Estimated Δz	V_m m/s	$\Delta V = 2(V_m - V_i)$	$15\!\cdot\!3(\Delta V/\Delta t)$	$0\!\cdot\!255\,V_m{}^2$	z_m from eqn. (8)	z_m from eqn. (10)
0	5	−1·030	2·01	Using a 5sec interval, $V_m = 1\!\cdot\!8\Delta z$ and $15\!\cdot\!3(\Delta V/\Delta t) = 30\!\cdot\!6\Delta V$						
				1·10	1·980	−0·060	−0·184	1·000	−0·816	−0·48
				1·05	1·890	−0·240	−0·734	0·911	−0·177	−0·505
				1·075	1·935	−0·150	−0·459	0·955	−0·496	−0·493
5	5	+0·045	1·86							
				0·92	1·656	−0·408	−1·248	0·699	+0·549	+0·505
				0·924	1·663	−0·394	−1·204	0·705	+0·499	+0·507
				0·923	1·661	−0·397	−1·215	0·704	+0·511	+0·506
10	5	0·968	1·463							
				0·650	1·170	−0·586	−1·793	0·349	1·444	1·293
				0·660	1·188	−0·550	−1·683	0·360	1·323	1·298
				0·662	1·192	−0·543	−1·662	0·362	1·300	1·299
15	5	1·630	0·920							
				0·330	0·594	−0·652	−1·995	0·090	1·905	1·795
				0·340	0·612	−0·616	−1·885	0·096	1·789	1·800
				0·339	0·610	−0·620	−1·896	0·095	1·801	1·800
20	2	1·969	0·30	The peak level is close, reduce Δt to 2sec, $V_m = 4\!\cdot\!5\Delta z$, $15\!\cdot\!3(\Delta V/\Delta t) = 7\!\cdot\!65\Delta V$						
				0·040	0·180	−0·240	−1·836	0·008	1·828	1·989
				0·037	0·167	−0·267	−2·043	0·007	2·036	1·987
				0·038	0·171	−0·258	−1·974	0·007	1·987	1·988
22		2·007	0·042	This is effectively the peak level						

Maximum rise of water level = **2·007m** at 22 secs from closure

Problems

1 The outlet valve of a pipeline 1500 m long is closed in 1 s. If the initial velocity of flow of the water is $1\!\cdot\!2$ m/s, find the rise in pressure assuming the pipe to be rigid.
Answer 1800 kN/m²

2 A uniform pipe, 60 m long, is fitted with a plunger which is brought to rest uniformly in $1 \cdot 5$ s from an initial velocity of $1 \cdot 8$ m/s. Assuming the water to be incompressible determine the pressure on the piston caused by the retardation.

If instead of being retarded uniformly the plunger is driven by a crank, $0 \cdot 3$ m long and making 100 rev/min, so that the piston travels with simple harmonic motion, determine the maximum pressure due to the retardation of the piston.
Answer 72 kN/m², 1960 kN/m²

3 Show that the velocity of transmission of a pressure wave along a pipe carrying a compressible fluid is given by

$$v = \sqrt{\frac{1}{\rho\left(\dfrac{1}{K} + \dfrac{d}{tE}\right)}}$$

if any longitudinal strain in the pipe is disregarded, where ρ is density of the fluid, K = bulk modulus of fluid, d = diameter, t = thickness of pipe, E = elastic coefficient of material of pipe.

Describe the phenomenon of waterhammer such as occurs in a pipe when flow is suddenly stopped by closing a valve at the far end.

Water flows in a pipe 1200 m long and a valve at the far end is suddenly closed. How long will the rise of pressure at the valve persist if the elasticity of the pipe is negligible? K for water is 2070 GN/m².
Answer $1 \cdot 67$ s

4 Water flows at a steady velocity of 2 m/s in a pipe of length 500 m and of uniform circular cross-section. Discharge is to atmosphere through a valve and the static pressure at the valve under steady flow conditions is 300 m of water.

Calculate the ratio of the internal diameter d of the pipe to the wall thickness t so that the increase in hoop stress is limited to 2×10^7 N/m² upon complete and instantaneous closure of the valve. Bulk modulus of water = 2×10^9 N/m² and for the pipe E = 2×10^{11} N/m².
Answer $15 \cdot 18$

5 Calculate the maximum permissible discharge in a 150 mm pipe if the pressure is not to exceed 1400 kN/m² when the outlet is suddenly closed. For water $K = 2 \cdot 07 \times 10^9$ N/m² and the pipe may be assumed to be rigid.
Answer $0 \cdot 0172$ m³/s.

6 A cast iron pipe 150 mm diam and 15 mm thick is conveying water when the outlet is suddenly closed. Calculate the maximum permissible discharge if the pressure rise is not to exceed 1700 kN/m². Take K for water as $2 \cdot 14 \times 10^9$ N/m² and E for cast iron as 117×10^9 N/m². Ignore lateral strain effects.
Answer $22 \cdot 76$ dm³/s

7 If the pressure is not to exceed 700 kN/m² in a 200 mm pipe when the outlet is suddenly closed, calculate the maximum

permissible discharge if $K = 2 \cdot 07 \times 10^9$ N/m². Assume that the pipe is rigid.
Answer $0 \cdot 0153$ m³/s

8 A straight pipe 900 m long, 300 mm diam and 25 mm wall thickness conveys water to a turbine at a rate of $0 \cdot 17$ m³/s. If the turbine gates are suddenly completely closed, calculate the pressure rise in the pipeline by the strain energy theory.

Compare this with that obtained by multiplying the pressure rise as calculated from the momentum loss by the empirical factor $1 \cdot 33$, assuming constant retardation and that the gates are closed in $0 \cdot 9$ s. Work from first principles and take E for the pipe material as 124×10^9 N/m², $\sigma = 0 \cdot 25$, and K for the water as $2 \cdot 07 \times 10^9$ N/m².
Answer $3 \cdot 17 \times 10^6$ N/m², 1 : 102

9 Water from a pipeline having a length of 720 m is discharged through a needle-controlled nozzle. The total head measured at the entrance to the nozzle is 150 m when the steady velocity of flow through the pipeline is $2 \cdot 4$ m/s. The nozzle aperture which is initally equal to the area of the pipe is closed in 6 s by the needle moving so that the nozzle area decreases uniformly with time. Assuming that the water and the pipe are inelastic so that there is no pressure wave and that the discharge coefficient for the nozzle is constant, find the maximum increase in pressure head at the entrance to the nozzle. Work from first principles or prove any formula used.
Answer $32 \cdot 37$ m

10 Water from a reservoir flowing through a rigid 150 mm diam pipe with a velocity of $2 \cdot 5$ m/s is completely stopped by closure of a valve situated 1050 m from the reservoir. Determine the maximum rise of pressure in kN/m² above that corresponding to uniform flow, when valve closure takes place in (*a*) 1 s and (*b*) 5 s. Assume that the pressure increases at a uniform rate and that there is no damping of the pressure wave.

Give diagrams showing how the velocity in the pipe varies with time during the period of valve closure. The velocity of sound in water is 1420 m/s.
Answer 3550 kN/m², 1052 kN/m²

11 A pipline of 600 mm diam is 1500 m long. When the discharge is $0 \cdot 51$ m³/s the valve at the outlet end is closed in 10 s, in such a way that the retardation at any instant is proportional to the time which has elapsed since the commencement of the closure. Calculate the rise of pressure.

Find also the rise of pressure which would take place if the closure were practically instantaneous, taking the bulk modulus for water, modified to allow for the elasticity of the pipe, as $1 \cdot 9 \times 10^9$ N/m².
Answer 543 kN/m², 2488 kN/m²

12 Calculate the velocity of a pressure wave through a liquid in terms of the bulk modulus K and the mass density ρ. Describe how the pressure in a pipeline varies when the flow is suddenly stopped by the closure of a valve at the outlet end.

A pipe is 1500 m long and water flows through it with a velocity of 1·2 m/s. If the flow is stopped instantaneously by the closure of a valve at the outlet end, draw a graph showing the variation of pressure with time at the valve and at a point 600 m from the valve. Neglect friction and the elasticity of the pipe. $K = 2\cdot07 \times 10^9$ kN/m². Draw one complete cycle of pressure change.

Answer Pressure rise = $1\cdot73 \times 10^6$ N/m²

13 Describe the operation of a simple surge tank communicating with the pipeline supplying the turbines in a hydro-electric plant.

Show that if the friction head is proportional to (velocity)², the oscillatory motion of the level in the surge tank following sudden complete shut down of the turbines is given by an equation taking the form

$$\frac{d^2H}{dt^2} + A\left(\frac{dH}{dt}\right)^2 + BH = 0$$

in which H is the height at any instant of the surge tank level with reference to the reservoir level, A and B are constants, the former having positive value when the flow along the pipe-line is towards the surge tank and negative when reversed.

Find A and B if the surge tank diameter is 30 m, pipeline diameter 4·5 m, and the length of pipeline from reservoir to surge tank 720 m. At the instant when the turbines are completely shut down the flow along the pipeline is 42·4 m³/s and the level in the surge tank is stationary, 0·9 m below the level in the reservoir.

Answer $A = 0\cdot077, B = 3\cdot07 \times 10^{-4}$

14 A low pressure tunnel 2400 m long with a cross-sectional area of 5·4 m² feeds water, via a simple surge shaft of cross-sectional area 108 m² to the high pressure penstock of a hydro-electric power house.

Initially the steady rate of flow to the turbines is 12·15 m³/s. A valve on the high pressure penstock is completely closed in a period of 5 min in such a way as to produce a linear rate of reduction of discharge with time. Neglecting friction, calculate the maximum rise of the water level in the surge shaft. (The solution of the equation $d^2z/dt^2 + m^2z = c^2$ is $z = A \sin mt + B \cos mt + c^2/m^2$.)

Answer   3·68 m, 220 s after start

15 In a hydro-electric scheme the water velocity in the low pressure penstock under steady full load conditions is 3·3 m/s. The low pressure penstock is 970 m long, 2 m diam, with $f = 0\cdot005$ and connects to the base of a simple surge tank 6 m diam. The connection to the base of the surge tank is 21 m below reservoir surface level. When starting the turbines, the gates are suddenly completely opened to the steady full-load flow. Estimate the minimum depth of water in the surge tank due to mass oscillation. Use the basic equations of mass oscillation to form a step-by-step integration.

Answer 10·3 m

16 A simple hydro-electric scheme has a low-pressure concrete-lined tunnel, a simple surge shaft and a high pressure penstock to a reaction turbine which discharges into a tailrace tunnel.

State the purpose of surge shafts (and/or chambers) in such schemes and explain with diagrams and sketched graphs the criteria involved in determining the limits of depth and height for the simple upstream surge shaft.

Explain the economic disadvantages associated with the use of the simple surge shaft; discuss other types, and give the advantages to be gained from using them.

Develop the basic equations of mass oscillation for a simple surge shaft with tunnel friction together with expressions for the periodic time and maximum positive or negative surge for the frictionless case.

11

Non-uniform flow in channels

In considering uniform flow in channels in Volume I it was assumed that successive cross-sections and corresponding mean velocities were everywhere the same and that the loss of head in friction was equal to the fall of the channel bed so that bed, water surface and energy gradient were parallel.

In non-uniform flow none of these conditions need apply. Depth may vary from section to section and the energy gradient, water surface and bed need no longer be parallel. Loss of head is measured by the fall in the energy gradient and not by the fall in the water surface or the bed level. If the flow is accelerating the slope of the bed is greater than that of the energy gradient.

Specific energy and critical depth

11.1 Specific energy

> Define the term "specific energy" for flow in a rectangular channel and show that the depth of flow varies with specific energy and discharge.

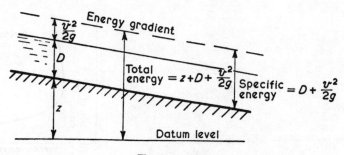

Figure   11.1

Solution. Under steady flow conditions the energy gradient is parallel to the bed of the channel (Fig. 11.1) and it is often convenient to consider the flow as if it were frictionless and the bed were horizontal. The specific energy H is the energy of the fluid reckoned above bed level and is measured as a head.

$$H = D + \frac{v^2}{2g}$$

Thus   Mean velocity $v = \sqrt{[2g(H - D)]}$

If $B =$ width of the channel

Discharge $Q = BDv = BD\sqrt{[2g(H - D)]}$

or $D^3 - HD^2 \times \dfrac{Q^2}{2gB^2} = 0$

This equation has three roots of which two are positive and the other is negative and unreal.

Figure 11.2 shows that variation of Q/B with depth D when the

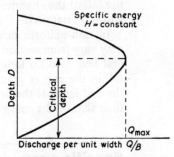

Figure 11.2

specific energy H is constant. For a given value of H there will be two alternative depths for a given discharge. Similarly for a given value of discharge Q there will be two, and only two, alternative depths for a given specific energy as shown in Fig. 11.3. In both cases it can be seen

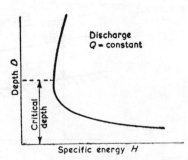

Figure 11.3

that there is a "critical depth" at which the two roots coincide and at which the energy required for a given discharge is a minimum and the discharge for a given energy is a maximum.

11.2 Critical velocity and critical depth

Define the term "critical velocity" and derive an expression for the critical velocity in any channel in terms of the discharge Q, area of cross-section A, and width of water surface B. Hence show that in a rectangular channel the critical depth is two-thirds of the specific energy H and that the Froude number for critical depth conditions is unity.

Water flows in a channel of rectangular section with a velocity of $1 \cdot 5$ m/s and a depth of $1 \cdot 2$ m. Determine (a) the specific energy of the flow, (b) the critical depth, (c) the maximum discharge under critical flow conditions if the channel is 3 m wide.

Solution. The critical velocity is the velocity in a channel when the fluid flows at critical depth.

Referring to Fig. 11.4, for any depth D

Specific energy
$$H = \frac{v^2}{2g} + D$$

or since $v = Q/A$,
$$H = \frac{Q^2}{2A^2g} + D$$

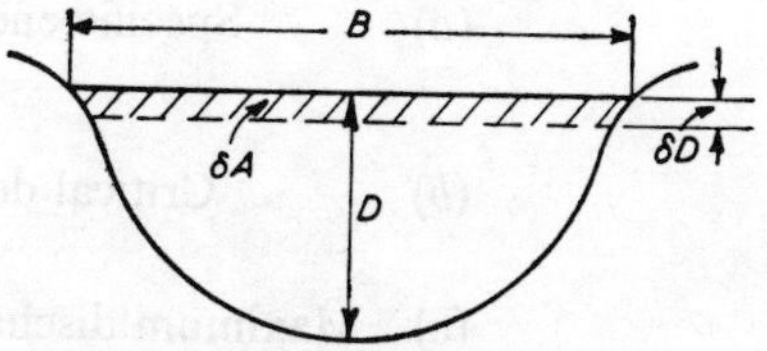

Figure 11.4

For flow at critical depth and velocity the specific energy is a minimum for a given value of Q (example **11.1**), and

$$\frac{dH}{dD} = 0$$

or
$$-\frac{2Q^2A^{-3}}{2g}\frac{dA}{dD} + 1 = 0$$

From Fig. 11.4
$$\delta A = B\delta D, \quad \frac{dA}{dD} = B$$

Substituting
$$\frac{Q^2B}{A^3g} = 1$$

for critical flow condition and if v_c is the critical velocity

$$v_c = \frac{Q}{A}$$

so that
$$\frac{v_c^2 B}{Ag} = 1 \quad \text{or} \quad v_c = \sqrt{\frac{Ag}{B}}$$

For a rectangular channel. $A = BD_c$ where D_c is the critical depth. Therefore

$$v_c = \sqrt{(D_c g)} \qquad (1)$$

Specific energy $H = \dfrac{v_c^2}{2g} + D_c = \dfrac{3}{2} D_c$

Critical depth $D_c = \frac{2}{3} H$

Also from equation (1)

$$\text{Froude number} = \frac{v_c}{\sqrt{(gD_c)}} = 1$$

Note also that since $Q^2 B / A^3 g = 1$ and $A = BD$

$$\text{Critical depth } D_c = \sqrt[3]{\frac{Q^2}{gB^2}}$$

and since $v_c = \sqrt{(gD_c)}$

$$v_c = \sqrt[3]{\frac{Qg}{B}}$$

(a) $\text{Specific energy} = \dfrac{v^2}{2g} + D = \dfrac{1\cdot 5^2}{2 \times 9\cdot 81} + 1\cdot 2 = \mathbf{1\cdot 315\,m}$

(b) $\text{Critical depth} = \dfrac{2}{3}\left(\dfrac{v^2}{2g} + D\right) = \dfrac{2}{3} \times 1\cdot 315 = \mathbf{0\cdot 875\,m}$

(c) $\text{Maximum discharge} = BD_c v_c = BD_c \sqrt{[2g(H - D_c)]}$

$$= 3 \times 0\cdot 875\sqrt{[2 \times 9\cdot 81(1\cdot 315 - 0\cdot 875)]}$$

$$= \mathbf{7\cdot 73\,m^3/s}$$

11.3 Shooting flow, tranquil flow, critical depth in triangular channel

Explain the meaning of the terms "shooting flow" and "tranquil or streaming flow" in connexion with open channels.

Deduce an expression for the critical depth of flow in a channel of triangular section, with side slopes of 1 vertical to n horizontal, in terms of the specific energy H.

Solution. "Shooting flow" is the term applied to the shallow fast flow which occurs when the depth in the channel is less than the critical depth.

"Tranquil or streaming flow" is the term for the deep slow flow at depths above the critical depth.

Specific energy $\quad H = D + \dfrac{v^2}{2g}$

$$v = \sqrt{[2g(H - D)]}$$

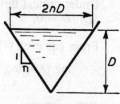

Figure   11.5

Area of section $A = \frac{1}{2}D \times 2nD$ from Fig. 11.5

$$= nD^2$$

Discharge $Q = nD^2\sqrt{[2g(H - D)]}$ (1)

At critical depth, when H is a constant, Q is a maximum and $dQ/dD = 0$.

From (1), $\log Q = \log n + 2 \log D + \frac{1}{2} \log 2g + \frac{1}{2} \log (H - D)$
Differentiating

$$\frac{1}{Q}\frac{dQ}{dD} = \frac{2}{D} - \frac{1}{2(H - D)} = 0$$

$$4H - 5D = 0$$

$$\text{Critical depth} = \tfrac{4}{5}H$$

11.4   Critical depth in trapezoidal channel

Derive an expression for the critical depth in a trapezoidal channel. A channel of trapezoidal cross-section, width of base 0.6 m and side slopes $45°$ carries 0.34 m^3/s. Determine the critical depth.

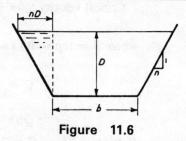

Figure   11.6

Solution. Consider a trapezoidal channel with side slopes 1 to n, Fig. 11.6, and bottom width b.

Area of section $= A = (b + nD)D$

Specific energy $= H = \dfrac{v^2}{2g} + D$

$$v = \sqrt{[2g(H - D)]}$$

Discharge $Q = Av = D(b + nD)\sqrt{[2g(H - D)]}$ (1)

Under critical flow conditions, for constant H, Q is a maximum, and $dQ/dD = 0$.

From equation (1),

$$\log Q = \log D + \log (b + nD) + \tfrac{1}{2}\log 2g + \tfrac{1}{2}\log (H - D)$$

Differentiating, $\qquad \dfrac{1}{Q}\dfrac{dQ}{dD} = \dfrac{1}{D} + \dfrac{n}{b + nD} + \dfrac{-1}{2(H - D)}$

Since $dQ/dD = 0$ when D = critical depth $= D_c$

$$\frac{1}{D_c} + \frac{n}{b + nD_c} - \frac{1}{2(H - D_c)} = 0$$

$$5nD_c^2 + (3b - 4nH)D_c - 2bH = 0 \qquad (2)$$

If n is not zero

$$D_c = -\frac{3b + 4nH + \sqrt{(9b^2 - 24bnH + 16n^2H^2 + 40bnH)}}{10n}$$

$$D_c = \frac{4nH - 3b + \sqrt{(16n^2H^2 + 16nHb + 9b^2)}}{10n} \qquad (3)$$

Note. This is a general solution. If $b = 0$ the channel is triangular and equation (3) gives $D_c = \tfrac{4}{5}H$. If $n = 0$ equation (3) does not apply but equation (2) gives $D_c = \tfrac{2}{3}H$ for a rectangular section.

In the present problem it is simpler to work from the expression for critical velocity in example **11.2**, since H is not known and depends on D_c.

Critical velocity $v_c = \dfrac{Q}{A_c} = \sqrt{\dfrac{gA_c}{B}}$

where B = top width and b = bottom width.

$$\therefore \qquad Q = \sqrt{\frac{A_c^3 g}{B}} = \sqrt{\frac{(b + nD_c)^3 D_c^3 g}{b + 2nD_c}}$$

$$Q^2(b + 2nD_c) = (b + nD_c)^3 D_c^3 g$$

Putting $\qquad Q = 0{\cdot}34\,\mathrm{m^3/s}, \quad b = 0{\cdot}6\,\mathrm{m}, \quad n = 1,$

$$0{\cdot}34^2(0{\cdot}6 + 2D_c) = (0{\cdot}6 + D_c)^3 D_c^3 \times 9{\cdot}81$$

$$D_c^6 + 1{\cdot}8D_c^5 + 1{\cdot}08D_c^4 + 0{\cdot}216D_c^3 - 0{\cdot}0236D_c - 0{\cdot}00708 = 0$$

Solving by trial,

Critical depth $D_c = \mathbf{0{\cdot}264\,m}$

11.5 Hydraulic jump

What is a "hydraulic jump" and under what conditions may it occur?

(*a*) Water issuing from a sluice enters a horizontal rectangular channel with uniform velocity v and depth of flow D. Show that a "hydraulic jump" or "standing wave" will be formed if the velocity v exceeds $\sqrt{(gD)}$.

(*b*) If the velocity when the water enters the channel is 3 m per sec and the depth of flow is 150 mm, obtain, working from first principles or proving any formula used:

(i) the depth of flow immediately after the jump;

(ii) the loss of specific energy due to the formation of the jump.

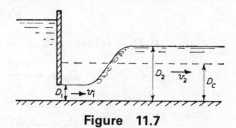

Figure   11.7

Solution. If flow in a channel cannot be maintained below the critical depth the change to tranquil flow above the critical depth occurs suddenly by means of a hydraulic jump (Fig. 11.7). A jump will occur if the downstream water level is raised by an obstruction above the critical depth. Alternatively since the frictional loss depends on the velocity it will be greater for shooting flow than for tranquil flow, so that if the bed slope is insufficient to overcome the frictional loss for shooting flow, a transition to tranquil flow will occur by means of a jump.

(*a*) Since the channel is horizontal the bed slope is insufficient to maintain flow against frictional resistance. A hydraulic jump will therefore occur provided that the depth of the flow leaving the sluice is less than critical depth so that $D_c > D$.

Now from example **11.2** for a rectangular channel

$$D_c = \frac{2}{3} H = \frac{2}{3}\left(D_c + \frac{v^2}{2g}\right) = \frac{v^2}{g}$$

A jump will therefore occur when $\dfrac{v^2}{g} > D$ or v exceeds $\sqrt{(gD)}$.

(*b*) When a jump occurs there is a change of momentum since the flow is slowed down. The force producing this change is due to the difference in hydrostatic pressures resulting from the change of depth.

(i)   Rate of change of momentum $= \rho Q\,(v_1 - v_2)$

If the width of channel is B

Hydrostatic force acting upstream $= \frac{1}{2}\rho g D_2{}^2 B$

$$\text{Hydrostatic force acting downstream} = \tfrac{1}{2}\rho g D_1{}^2 B$$

$$\text{Resultant force} = \tfrac{1}{2}\rho g B(D_2{}^2 - D_1{}^2)$$

Also
$$Q = B D_1 v_1 = B D_2 v_2$$

Thus
$$\tfrac{1}{2}\rho g B(D_2{}^2 - D_1{}^2) = \rho B v_1 D_1 \left(v_1 - v_1 \frac{D_1}{D_2}\right)$$

$$(D_2{}^2 - D_1{}^2) = \frac{2v_1{}^2 D_1}{g D_2}(D_2 - D_1)$$

$$v_1{}^2 = \frac{g D_2}{2 D_1}(D_2 + D_1)$$

$$D_2 = -\frac{D_1}{2} + \sqrt{\left(\frac{2v_1{}^2 D_1}{g} + \frac{D_1{}^2}{4}\right)}$$

If $D_1 = 0\cdot15\,\text{m}$ and $v_1 = 3\,\text{m/s}$

$$D_2 = -\frac{0\cdot15}{2} + \sqrt{\left(\frac{2 \times 9 \times 0\cdot15}{g} + \frac{0\cdot15^2}{4}\right)}$$

$$= -0\cdot075 + \sqrt{(0\cdot276 + 0\cdot0056)}$$

$$D_2 = -0\cdot075 + 0\cdot531 = \mathbf{0\cdot456\,m}$$

(ii) For continuity of flow $D_1 v_1 = D_2 v_2$

$$v_2 = v_1 \frac{D_1}{D_2} = 3 \times \frac{0\cdot15}{0\cdot456} = 0\cdot985\,\text{m/s}$$

$$\text{Loss of specific energy} = \left(D_1 + \frac{v_1{}^2}{2g}\right) - \left(D_2 + \frac{v_2{}^2}{2g}\right)$$

$$= 0\cdot15 + \frac{9}{2 \times 9\cdot81} - \left(0\cdot456 + \frac{0\cdot985^2}{2 \times 9\cdot81}\right)$$

$$= 0\cdot15 + 0\cdot459 - 0\cdot456 - 0\cdot0494$$

$$= \mathbf{0\cdot104\,m}$$

Venturi flume, standing wave flume, and broad-crested weir

11.6 Venturi flume

(a) Derive an expression for the rate of flow through a venturi flume and show that this will be a maximum when the depth at the throat is equal to the critical depth.

(b) A venturi flume is $1\cdot2$ m wide at entrance and $0\cdot6$ m in the throat; the bottom is horizontal. Neglecting hydraulic losses in the flume calculate the flow if the depth at entrance and throat is $0\cdot6$ m and $0\cdot55$ m respectively.

(c) A hump is now installed at the throat of height $0\cdot2$ m so that a standing wave is formed beyond the throat. Assuming the same flow as before, show that the increase in the upstream depth is nearly 10 cm. Work from first principles or prove any formula used.

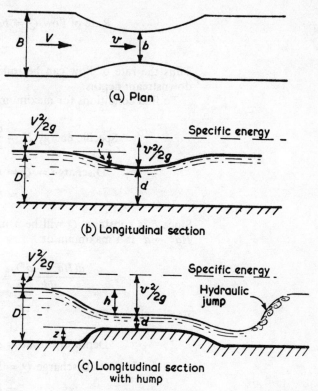

(a) Plan

(b) Longitudinal section

(c) Longitudinal section
with hump

Figure 11.8

Solution (a) A constriction in a channel forms a venturi flume and can be used for flow measurement. As shown in Fig. 11.8 the width of the channel is reduced or the floor raised by means of a hump (Fig. 11.8(c)). Since the specific energy is constant the increased velocity in the throat is accompanied by a fall in the level of the surface h.

Referring to Fig. 11.8(b), since the specific energy upstream and at the throat is constant

$$D + \frac{V^2}{2g} = d + \frac{v^2}{2g} \qquad (1)$$

For continuity of flow $\qquad BDV = bdv$

$$V = \frac{bd}{BD} v$$

Substituting in equation (1)

$$\frac{v^2}{2g}\left(1 - \frac{b^2 d^2}{B^2 D^2}\right) = D - d = h$$

$$v = \frac{\sqrt{(2gh)}}{\sqrt{\left(1 - \frac{b^2 d^2}{B^2 D^2}\right)}}$$

$$\text{Rate of flow } Q = bdv = bd\sqrt{\dfrac{2gh}{1 - \left(\dfrac{bd}{BD}\right)^2}} \qquad (2)$$

Thus the rate of flow can be found by measuring the upstream and downstream depths.

To find conditions for maximum flow:

$$\text{Specific energy } H = \frac{v^2}{2g} + d \qquad (3)$$

$$\text{Discharge} = Q = bdv = bd\sqrt{[2g(H - d)]}$$
$$= b\sqrt{(2g)}(Hd^2 - d^3)^{1/2}$$

Since b is constant, Q will be a maximum for a given value H when $Hd^2 - d^3$ is a maximum or

$$\frac{d(Hd^2 - d^3)}{dd} = 2Hd - 3d^2 = 0$$

For maximum flow

$$\text{Depth of throat} = d = \tfrac{2}{3}H = \text{critical depth}$$
$$\text{Discharge } Q = bd\sqrt{[2g(H - d)]}$$
$$= b \times \tfrac{2}{3}H\sqrt{[2g \times \tfrac{1}{3}H]} = 1{\cdot}706bH^{3/2}$$
$$(4)$$

Thus the rate of flow can be found from the upstream depth and the geometry of the flume.

The existence of maximum flow conditions can be verified by arranging for a standing wave to form downstream. This arrangement is known as a *standing wave flume*.

(b) For a venturi flume from equation (2)

$$Q = bd\sqrt{\dfrac{2gh}{1 - \left(\dfrac{bd}{BD}\right)^2}}$$

Putting $b = 0{\cdot}6\,\text{m}$, $B = 1{\cdot}2\,\text{m}$, $d = 0{\cdot}55\,\text{m}$, $D = 0{\cdot}6\,\text{m}$ and since the floor is level

$$h = D - d = 0{\cdot}05\,\text{m}$$

$$Q = 0{\cdot}6 \times 0{\cdot}55\sqrt{\dfrac{2 \times 9{\cdot}81 \times 0{\cdot}05}{1 - \left(\dfrac{0{\cdot}6 \times 0{\cdot}55}{1{\cdot}2 \times 0{\cdot}6}\right)^2}} = \mathbf{0{\cdot}368\,m^3/s}$$

(c) When a standing wave is formed (Fig. 11.8(c)) the conditions are those of maximum flow and the depth $d = \tfrac{2}{3}$ of the specific energy at the throat. Since in equation (3) H was taken as $v^2/2g + d$ the specific energy will be measured above bed level at the throat which is the top of the hump.

$$Q = 1.706bH^{3/2}$$

$$= 1.706b \left(\frac{v^2}{2g} + d \right)^{3/2}$$

Now by Bernoulli's equation

$$\frac{v^2}{2g} + d = \frac{V^2}{2g} + (D - Z)$$

$$Q = 1.706b \left[\frac{V^2}{2g} + (D - Z) \right]^{3/2}$$

$$V = \frac{Q}{BD}$$

Putting $Q = 0.368 \, \text{m}^3/\text{s}$ and, assuming a rise of $10 \, \text{cm} = 0.1 \, \text{m}$ upstream, $D = 0.7 \, \text{m}$

$$V = \frac{0.368}{1.2 \times 0.7} = 0.429 \, \text{m/s}, \quad \frac{V^2}{2g} = \frac{0.429^2}{19.62} = 0.0094 \, \text{m}$$

Thus

$$Q = 1.706b\{0.0094 + (D - z)\}^{3/2}$$

Putting $Q = 0.368 \, \text{m}^3/\text{s}$, $b = 0.6 \, \text{m}$, $z = 0.2 \, \text{m}$

$$\{0.0094 + (D - 0.2)\}^{3/2} = \frac{0.368}{1.706 \times 0.6} = 0.3595$$

$$0.0094 + D - 0.2 = 0.5056$$

$$\text{Upstream depth} = D = 0.6962 \, \text{m}$$

$$\text{Increase in depth} = 0.0962 \, \text{m} \simeq \mathbf{10 \, cm}$$

11.7 Venturi flume

A venturi flume in a rectangular channel of width B has a throat width of b. The depth of liquid at entry is H and at the throat is h. Derive an expression for the theoretical volumetric flow rate of the liquid in terms of H, B and the ratio h/H and develop a relationship between the ratio h/H and the ratio b/B. State what assumptions you make regarding the downstream flow.

Solution. Assume that the flume is operating under maximum flow conditions with critical depth in the throat and a standing wave downstream (Fig. 11.9).

$$\text{Depth at throat } h = \tfrac{2}{3} \times \text{specific energy}$$

$$= \tfrac{2}{3} \left(\frac{v^2}{2g} + h \right)$$

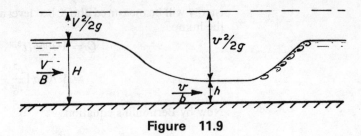

Figure 11.9

$$\frac{v^2}{2g} = \tfrac{1}{2}h$$

By Bernoulli's equation $\quad H + \dfrac{V^2}{2g} = h + \dfrac{v^2}{2g} = \dfrac{3}{2}h$

$$V = \sqrt{[2g(\tfrac{3}{2}h - H)]}$$

$$Q = BHV = BH^{3/2}\sqrt{\left[2g\left(\frac{3}{2}\frac{h}{H} - 1\right)\right]}$$

For continuity of flow $\quad BHV = bhv$

$$V = \frac{bh}{BH}v$$

or, since $\dfrac{v^2}{2g} = \tfrac{1}{2}h$ $\qquad \dfrac{V^2}{2g} = \dfrac{b^2h^2}{B^2H^2}\dfrac{h}{2}$

By Bernoulli's theorem $\quad H + \dfrac{V^2}{2g} = h + \dfrac{v^2}{2g}$

$$H + \frac{b^2h^2}{B^2H^2}\frac{h}{2} = h + \frac{h}{2} = \frac{3}{2}h$$

Dividing by $h/2$, $\quad 2\dfrac{H}{h} + \left(\dfrac{b}{B}\right)^2\left(\dfrac{h}{H}\right)^2 = 3$

$$\left(\frac{b}{B}\right)^2 = 3\left(\frac{H}{h}\right)^2 - 2\left(\frac{H}{h}\right)^3$$

11.8 Broad-crested weir

Deduce from first principles an expression for the discharge over a flat-topped broad-crested weir forming a spillway to a large reservoir.

Calculate the discharge over a broad-crested weir 15 m wide when the upstream level is 0·6 m above the crest and the coefficient of discharge is 0·6.

Solution. Let H be the height of the reservoir level above the weir, h the depth on the crest of the weir and v the velocity (Fig. 11.10).

If the reservoir is large the velocity upstream of the weir is negligible and by Bernoulli's theorem

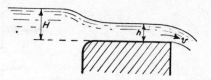

Figure 11.10

$$H = h + \frac{v^2}{2g}$$

$$v = \sqrt{[2g(H - h)]}$$

If L = width of weir (longitudinally) and C_d = coefficient of discharge

$$\text{Discharge} = Q = C_d L h v$$

$$Q = C_d L \sqrt{(2g)} (Hh^2 - h^3)^{1/2}$$

If the flow falls freely at the downstream edge the depth over the rest will be that which will give maximum discharge. For Q to be a maximum $(Hh^2 - h^3)$ must be a maximum or

$$\frac{d(Hh^2 - h^3)}{dh} = 0$$

$$2Hh - 3h^2 = 0, \quad h = \tfrac{2}{3}H$$

Thus

$$Q = C_d L \sqrt{(2g)} \tfrac{2}{3} H \sqrt{(\tfrac{1}{3}H)}$$

$$= C_d \times 1 \cdot 706 L H^{3/2}$$

Putting $L = 15\,\text{m}, \quad C_d = 0 \cdot 6, \quad H = 0 \cdot 6\,\text{m}$

$$Q = 0 \cdot 6 \times 1 \cdot 706 \times 15 \times 0 \cdot 6^{3/2}$$

$$= 7 \cdot 13\,\text{m}^3/\text{s}$$

11.9 Water surface profile

Show that in an open channel of constant width, the slope of the water surface with respect to the bed is given by

$$\frac{dh}{dL} = \frac{i - j}{1 - (v^2/gh)}$$

where h = depth of flow, i = bed slope, j = friction loss per unit length, v = velocity of flow, and hence explain the distinction between uniform and non-uniform flow. What will be the actual slope of the water surface in millimetres per kilometre at a section in a rectangular channel at which the discharge is 28 m³/s, width 15 m, depth 3 m, bed slope 1 in 6700 and C (in the Chezy expression) = 61?

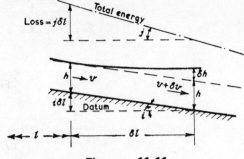

Figure 11.11

Solution. If the velocity changes by δv and the depth by δh in a length δl (Fig. 11.11) then by Bernoulli's equation, taking the datum shown

$$\left(i\delta l + h + \frac{v^2}{2g}\right) = \left(h + \delta h + \frac{(v + \delta v)^2}{2g}\right) + j\delta l$$

Ignoring products of small quantities

$$i\delta l = \delta h + \frac{v\delta v}{g} + j\delta l$$

so that

$$\frac{dh}{dl} = i - j - \frac{v}{g}\frac{dv}{dl} \qquad (1)$$

Rate of flow per unit width

$$q = vh = \text{constant}$$

or

$$\frac{d(vh)}{dl} = 0, \quad v\frac{dh}{dl} + h\frac{dv}{dl} = 0$$

$$\frac{dv}{dl} = -\frac{v}{h}\frac{dh}{dl}$$

Substituting in (1)

$$\frac{dh}{dl} = i - j + \frac{v^2}{gh}\frac{dh}{dl}$$

$$\frac{dh}{dl} = \frac{i - j}{1 - (v^2/gh)}$$

Note that in uniform flow the depth is constant, $dh/dl = 0$, and therefore the energy gradient is parallel to the bed since $i = j$.

$$\text{Mean velocity} = v = \frac{Q}{Bh} = \frac{28}{15 \times 3} = 0.623\,\text{m/s}$$

If j is the slope of the energy gradient, by the Chezy formula

$$v = C\sqrt{(mj)}$$

$$\text{Hydraulic mean depth} = m = \frac{\text{area of flow}}{\text{wetted perimeter}}$$

$$= \frac{Bh}{B + 2h} = \frac{15 \times 3}{15 + 6} = 2 \cdot 142 \, \text{m}$$

$$i = \frac{v^2}{C^2 m} = \frac{0 \cdot 623^2}{61^2 \times 2 \cdot 142} = 0 \cdot 487 \times 10^{-4}$$

$$\text{Bed slope} = i = 1/6700 = 1 \cdot 492 \times 10^{-4}$$

$$\frac{dh}{dl} = \frac{i - j}{1 - \frac{v^2}{gh}} = \frac{(1 \cdot 492 - 0 \cdot 487) \times 10^{-4}}{1 - \frac{0 \cdot 623^2}{g \times 3}}$$

$$= \frac{1 \cdot 005 \times 10^{-4}}{0 \cdot 988 \, 62} \, \text{m/m}$$

$$= \mathbf{101 \cdot 6 \, mm/km}$$

11.10 Backwater curve

(a) $8 \cdot 5$ m³/s of water flow in a channel of rectangular cross-section 6 m wide with a slope of 1 in 2000 and a depth of $0 \cdot 9$ m. Calculate the value of C in the Chezy formula.

(b) In order to raise the depth of water upstream a dam is built downstream. If the depth of water at the dam is $1 \cdot 8$ m, calculate how far upstream the depth will be $1 \cdot 5$ m.

(b) In order to raise the depth of water upstream a dam is built downstream. If the depth of water at the dam is $1 \cdot 8$ m, calculate how far upstream the depth will be $1 \cdot 5$ m.

Solution. (a) Mean velocity $= v = \dfrac{\text{discharge}}{\text{area of flow}} = \dfrac{8 \cdot 5}{6 \times 0 \cdot 9}$

$$= 1 \cdot 576 \, \text{m/s}$$

The Chezy formula is $v = C\sqrt{(mi)}$

For uniform flow $\quad i = \text{bed slope} = 1/2000 = 0 \cdot 0005$

$$m = \text{hydraulic mean depth} = \frac{\text{area}}{\text{wetted perimeter}}$$

$$= \frac{6 \times 0 \cdot 9}{6 + 2 \times 0 \cdot 9} = \frac{5 \cdot 4}{7 \cdot 8} = 0 \cdot 694 \, \text{m}$$

$$C = \frac{v}{\sqrt{(mi)}} = \frac{1 \cdot 576}{\sqrt{(0 \cdot 694 \times 0 \cdot 0005)}} = \mathbf{84 \cdot 9}$$

(b) The solution is found by calculating the slope of the water surface at the mean depth of $1 \cdot 65$ m and hence the distance for the required change of $0 \cdot 3$ m in depth.

At the mean depth $\quad m = \dfrac{6 \times 1 \cdot 65}{6 + 2 \times 1 \cdot 65} = 1 \cdot 065 \, \text{m}$

Mean velocity $= v = \dfrac{8 \cdot 5}{6 \times 1 \cdot 65} = 0 \cdot 946 \, \text{m/s}$

$$\text{Bed slope} = i = 0{\cdot}0005$$

For non-uniform flow, by the Chezy formula

$$v = C\sqrt{(mj)}$$

where j = slope of energy gradient.

$$j = \frac{v^2}{C^2 m} = \frac{0{\cdot}946^2}{84{\cdot}9^2 \times 1{\cdot}065} = 1{\cdot}165 \times 10^{-4}$$

From example **11.9**

$$\text{Slope of surface} = \frac{dh}{dl} = \frac{i - j}{1 - (v^2/gh)}$$

$$= \frac{5 \times 10^{-4} - 1{\cdot}165 \times 10^{-4}}{1 - \dfrac{0{\cdot}946^2}{9{\cdot}81 \times 1{\cdot}65}} = 4{\cdot}06 \times 10^{-4}\,\text{m/m}$$

$$= \frac{\text{change of depth}}{\text{distance}}$$

For a change of level of $0{\cdot}3\,\text{m}$

$$\text{Distance} = \frac{0{\cdot}3}{4{\cdot}06 \times 10^{-4}} = \mathbf{738\,m}$$

Problems

1 Water in a channel has a velocity of $2{\cdot}7$ m/s when flowing at the critical depth. Plot a graph showing the discharge per unit width for various depths of flow greater and less than the critical depth, if the specific energy is constant and the channel is rectangular in cross-section. What will be the depth at which the discharge will be one-half as great as for flow at critical depth?
Answer $0{\cdot}24$ or $1{\cdot}06$ m

2 In a rectangular channel the specific energy of flow is $1{\cdot}8$ J/N. Determine the critical depth and rate of discharge per unit width at this depth. If the discharge remains constant plot a graph showing the variation of specific energy with depth and determine its value when the depth of flow is $1{\cdot}8$ m.
Answer $1{\cdot}2$ m, $4{\cdot}12$ m³/s/m, $2{\cdot}07$ J/N

3 Water flows in a channel of trapezoidal section with a velocity of $4{\cdot}5$ m/s and a depth of $0{\cdot}6$ m at the centre. If the base width is $1{\cdot}2$ m and the side slopes are 1 vertical and 2 horizontal, determine (*a*) the specific energy, (*b*) the critical depth, (*c*) the discharge under critical flow conditions.
Answer $1{\cdot}63$ J/N, $1{\cdot}26$ m, $12{\cdot}63$ m³/s

4 Explain clearly with diagrams what you mean by "critical depth of flow" in a channel.

A wide rectangular channel is set to give a uniform depth of flow equal to that at the open end. Show that this slope is $\frac{1}{2}f$ where f is the frictional coefficient.

A channel of trapezoidal section, 0·6 m wide at the base, and with sides set at 60 deg. to the horizontal, carries 28 m³/s of water. Find the depth for critical flow.

Answer   0·26 m

5   Derive a formula for the height of a hydraulic jump.

A jump occurs in a channel of rectangular cross-section 6 m wide in which the rate of flow is 22·6 m³/s. If the depth of water before the jump occurs is 0·45 m, determine: (*a*) the depth after the jump has taken place, (*b*) the specific energy of the water before and after the jump, (*c*) the loss of power due to the jump.

Answer   (*a*) 2·322 m; (*b*) 4·04 m, 2·46 m; (*c*) 351 kW

6   A rectangular channel 3·6 m wide carries 2·5 m³/s. What will be the critical depth?

If the depth is reduced to half the critical depth and conditions are such that a hydraulic jump is formed, find the depth after the jump.

Answer   0·366 m, 0·647 m

7   A long straight channel of rectangular section 3 m wide has a slope of 1 in 2000 and the Chezy coefficient C is 77. The depth of water for normal flow is 1·8 m and the partial lowering of a sluice gate, the bottom of which is horizontal, raises the level upstream of the sluice by 0·6 m.

Assuming that the flow remains unchanged, find how far back from the sluice the depth will be 2·25 m and the maximum height in metres of the opening beneath the sluice if a standing wave is to be formed downstream of the sluice.

Answer   608 m, 0·93 m

8   The flow of liquid in an open channel is 0·2 m³/s. An adjustable sluice gate spans the entire width of the channel which is 0·9 m. What depth of water downstream of the sluice is required to cause critical flow?

The opening is adjusted to cause a standing wave downstream, such that the depth just before the wave is 10 cm. Determine the loss of specific energy in J/N at the standing wave. Prove any formula used.

Answer   0·1715 m, 0·04 J/N

9   Define the terms "specific energy" and "critical depth" used in connexion with non-uniform flow along open channels and show that the critical depth for flow along a rectangular channel having constant width B metres is given by

$$\sqrt[3]{(Q^2/gB^2)}$$

where Q is the rate of flow in m³/s.

Two consecutive lengths AB and BC of a rectangular open channel have uniform but different slopes. Shooting flow occurs along AB with depth below the critical value, and streaming flow occurs along BC with depth above the critical value so that a hydraulic jump is formed on the downstream side of the point B. Show that if D_a is the depth immediately preceding and D_b is the depth immediately following the jump, then

$$2D_c^3 = D_a D_b (D_a + D_b).$$

10 Find the downstream height of a hydraulic jump occurring on a level bed when the upstream depth is $0 \cdot 9$ m and the upstream velocity is $10 \cdot 5$ m/s. What conditions are necessary for a jump to occur?
Answer $4 \cdot 07$ m

11 A submerged sharp-edged weir extends across the entire width of a rectangular channel, the edge of the weir being $0 \cdot 75$ m above the bed of the channel. At a certain rate of flow the depth of water on the upstream side of the weir is $1 \cdot 2$ m and on the downstream side it is $1 \cdot 02$ m. Estimate the depth of water on the downstream side if the edge of the weir is raised 30 mm. Assume that the depth on the upstream side remains unaltered and ignore the velocity of approach.
Answer $0 \cdot 964$ m

12 Find the discharge over a flat-topped broad-crested weir 30 m long with rounded entrance when the upstream level is $0 \cdot 67$ m above the crest. Deduce any formula used and point out the limitations of the treatment $C_d = 0 \cdot 66$.
Answer $18 \cdot 5$ m³/s

13 The bed of a river 18 m wide has a uniform slope of 1 in 15 000, and the Chezy constant C is 50. A dam across the river has a spillway along the top, in the form of a broad-crested weir $16 \cdot 5$ m wide, the height of the sill being $3 \cdot 6$ m from the river bed.

Assuming that the whole of the river flow passes over the spillway, which has a coefficient of discharge of $0 \cdot 90$, find the afflux at the dam (i.e. the increase in the river depth caused by the dam) when the depth of the river well upstream is $1 \cdot 2$ m. Neglect the effect of the ends of the spillway, and assume that the banks of the river are vertical and of the same material as the bed. State fully any assumptions you make.
Answer $2 \cdot 843$ m

14 Two reservoirs A and B are joined by three 150 mm diam pipes 60 m long. Water flows from A to B and discharges over a broad-crested weir $0 \cdot 9$ m wide. The coefficient of discharge of the weir is $0 \cdot 85$, the depth over the weir being critical.

The sill of the weir is $1 \cdot 8$ m below the constant level in reservoir A. The pipe exit and entry are sharp and the friction coefficient is $0 \cdot 007$. Calculate the head over the weir and the quantity of water flowing.
Answer $0 \cdot 162$ m, $0 \cdot 085$ m³/s

15 A venturi flume is formed in a horizontal rectangular channel $1 \cdot 2$ m wide by constricting the channel to a width of $0 \cdot 9$ m and raising the floor level through the constriction by a height of

0·2 m. Calculate the rate of flow when the difference of level between throat and upstream is 25 mm and both up and downstream depths are 0·6 m. Calculate also the difference of level between upstream and throat when the flow is 0·45 m³/s and a standing wave is formed downstream.

Answer   0·268 m³/s, 0·129 m

16   The sides of a horizontal channel are contracted to form a venturi flume. The width of the channel is 1·2 m and of the throat is 0·6 m. The hydraulic jump on the downstream side ensures that the flow has maximum value. The depth of water on the upstream side is 0·6 m. Determine the discharge and depth of flow at the throat.

Answer   0·506 m³/s, 0·416 m

17   A venturi flume is to be installed in a channel conveying water with the object of raising the level of the water upstream. The channel is rectangular in section and is 12 m wide with a gradient of 1 in 6400 and a depth of water of 1·5 m. The width of the throat section is to be 6 m. If the bed of the flume at the throat is a streamline hump, find the necessary height of the hump in order that the depth of water upstream shall be 1·8 m. Take $v = 77\sqrt{(mi)}$, ignore losses in the flume and assume that a standing wave forms on the downstream side of the jump.

Answer   0·3326 m

18   A venturi flume is formed by contracting the sides of a channel, the bottom remaining level. If the depth of water just upstream is H and at the throat is h, derive from first principles an expression for the rate of flow through the flume if the upstream width is B and the throat width is b.

If in a venturi flume the width at entrance is 900 mm and at the throat 675 mm and the corresponding depths measured from a level bottom are 610 mm and 560 mm and the downstream level is also 610 mm, what will be the rate of flow through the flume?

Answer   0·516 m³/s

19   A channel of rectangular cross-section is 1·2 m wide and the normal flow is 0·7 m³/s with a depth of 0·6 m. If a streamline hump x m high is installed on the bed, draw a curve on squared paper showing how the theoretical ratio of the depth over the hump to the total energy immediately upstream varies with x. Take the datum for the energy as passing through the top of the hump.

Hence state whether a standing wave can form downstream when (a) $x = 0·1$ m and (b) when $x = 0·3$ m, giving reasons.

Answer   (a) No, (b) Yes

20   A venturi flume is placed in a channel 1·8 m wide in which the throat width is 1 m. The upstream depth is 0·9 m and the floor is effectively horizontal. Calculate the flow: (a) when the depth at the throat is 0·825 m; (b) when a standing wave is produced beyond the throat. Work from first principles or prove any formula used.

Answer   1·162 m³/s, 1·572 m³/s

21 Explain briefly how the depth upstream of a horizontal rectangular venturi flume varies with the width at the throat.

A venturi flume is constructed by contracting the sides of a rectangular channel having a width of 3 m and uniform slope of 1 in 2500. The rate of flow is that corresponding to a normal depth of $1 \cdot 8$ m. If the width of the throat is $1 \cdot 5$ m and a standing wave forms downstream, calculate the rate of flow in m^3/s, the depth immediately upstream of the flume and the depth at the throat. Take the Chezy coefficient $C = 66m^{1/6}$.

Answer $6 \cdot 24$ m^3/s, $1 \cdot 74$ m, $1 \cdot 21$ m

22 Prove that the rate of change of depth of water along an open channel of any shape is given by

$$\frac{dD}{dL} = \frac{(i - h_f)}{\left(1 + \frac{v}{g} \frac{dv}{dD} \right)}$$

where i is the slope of the channel invert, h_f is the loss of head per unit length and v is the mean velocity of flow.

In a horizontal rectangular channel having constant width of $1 \cdot 5$ m the depth of flow increases from $0 \cdot 9$ m to $0 \cdot 6$ m over a length of 150 m. Assuming that

$$h_f = \frac{0 \cdot 01 v^2}{2gm}$$

where m is the hydraulic mean depth, find the rate of flow in m^3/s. Use numerical integration.

Answer $1 \cdot 59$ m^3/s

23 Derive a formula for the rate of change of depth with length for a stream flowing in a rectangular channel.

Water is discharged into a rectangular channel $1 \cdot 2$ m wide by passing under a sluice so that the flow is $0 \cdot 85$ m^3/s and the depth $0 \cdot 6$ m. Examine how depth will vary downstream if the slope of the discharge channel is: (*a*) 1 in 1000; (*b*) 1 in 653; (*c*) 1 in 500. Assume a Chezy coefficient $= 55$ and illustrate your answer by sketches of the probable water surface.

Answer Depth (*a*) decreases, (*b*) constant, (*c*) increases

24 What do you understand by normal and critical flow conditions in a channel?

A wide channel of slope $i = 4/10\,000$ carries 93 dm^3/s of water per metre width. Given that k, the loss of friction per metre run, is expressed by $v = 55\sqrt{(mk)}$, determine: (*a*) the depth of water at the lower end, if this end of the channel is open; (*b*) the depth of water to be produced by a weir at the exit end, so that the normal flow shall occur; (*c*) how far upstream the depth will be $0 \cdot 210$ m if the depth at the weir is increased to $0 \cdot 215$ m assuming

$$\frac{dh}{dl} = \frac{i - k}{1 - (v^2/gh)}$$

Answer $0 \cdot 096$ m, $0 \cdot 193$ m, $40 \cdot 8$ m

25 A rectangular channel 1·2 m wide has a uniform slope of 1 in 1600 and normal flow depth is 0·6 m when the discharge is 0·57 m³/s. A sluice is lowered increasing the depth just upstream of it to 0·9 m. How far upstream of the sluice will the depth be 0·75 m?

Answer 423 m

12

Storage reservoir problems

Since the rainfall and the run-off from a catchment area vary from month to month and from year to year only the minimum yield can be relied upon unless a reservoir is provided to store the surplus when supply exceeds demand and to augment the flow when the supply is below demand. If the daily, weekly or monthly run-off from the catchment is known, the maximum permissible demand for water and the reservoir capacity required for any given demand can be found from the mass flow curve or by drawing a hydrograph or by tabulation.

12.1 Mass flow curves

The amounts of water flowing from a certain catchment area in each successive month (each assumed as 30 days) are given below in units of $1 \times 10^6 \mathrm{m}^3$:

$$2 \cdot 83, \ 3 \cdot 40, \ 5 \cdot 66, \ 18 \cdot 4, \ 23 \cdot 75, \ 23 \cdot 75, \ 20 \cdot 4, \ 9 \cdot 34,$$
$$7 \cdot 36, \ 6 \cdot 79, \ 6 \cdot 23, \ 5 \cdot 95$$

Determine
(a) The minimum capacity of a reservoir if the above water is to be drawn off at a uniform rate and none is to be lost by flow over the spillway.
(b) The amount of water which must be initially stored to maintain the above uniform draw-off.

Solution. (a) First plot the mass flow curve which is a graph showing total flow volume from the beginning of the period plotted as ordinates against time (*see* Table 12.1).

These figures are plotted in Fig. 12.1. If all the water is to be drawn off at a uniform rate, the graph of demand (volume drawn off) against time is the straight line OA.

Rate of uniform draw-off = slope of OA

$$= \frac{133 \cdot 86 \times 10^6}{30 \times 12} = 0 \cdot 371 \times 10^6 \mathrm{m}^3/\text{day}$$

From O to B the slope of the mass flow curve is less than that of the demand line indicating that the inflow to the reservoir is less than the outflow. At B the reservoir has reached its lowest level and since the slope of the mass flow curve now begins to exceed that of the demand line it will start to refill until at C the contents of the reservoir are the same as at O. The reservoir continues to fill until at D the inflow again

Table 12.1

Month	Monthly flow ($10^6 m^3$)	Total flow ($10^6 m^3$)
1	2·83	2·83
2	3·40	6·23
3	5·66	11·89
4	18·4	30·29
5	23·75	54·04
6	23·75	77·79
7	20·4	98·19
8	9·34	107·53
9	7·36	114·89
10	6·79	121·68
11	6·23	127·91
12	5·95	133·86

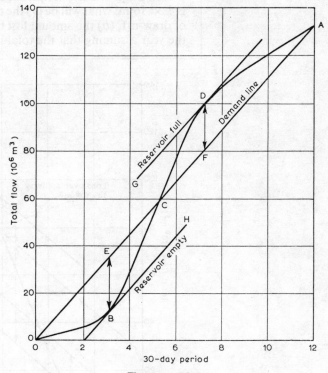

Figure   12.1

equals outflow after which point the reservoir begins to empty until at A the contents of the reservoir are again the same as at O. Thus BE represents the maximum volume withdrawn from the reservoir below its original level and DF is the maximum additional volume to be stored above the initial level.

Minimum capacity of reservoir required for the given demand OA is represented by (BE + EF).

$$\text{Minimum capacity} = \mathbf{41 \times 10^6 \, m^3}$$

This figure is conveniently found by drawing the tangents GD and BH to the mass flow curve parallel to OA and measuring the vertical distance between them.

(b) Since from O to B the inflow is less than the outflow the reservoir must contain sufficient to make good the difference. At B the reservoir is empty and the distance EB represents the excess of demand over supply to this point and therefore the amount of water which must initially be stored to maintain the uniform demand OA.

$$\text{Initial volume stored} = \mathbf{21{\cdot}5 \times 10^6 \, m^3}$$

12.2 Mass flow curves

If in the previous example the amount of water initially stored is $4 \times 10^6 \, m^3$ what will be (a) the maximum possible uniform rate of draw-off, (b) the amount lost by flow over the spillway during the year assuming that the total reservoir capacity is unaltered?

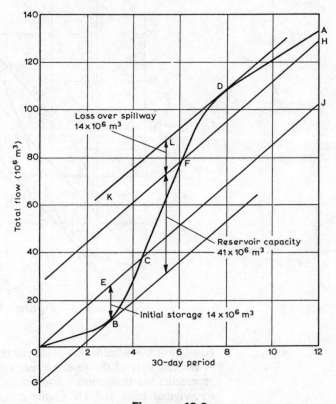

Figure 12.2

Solution (*a*) Using the same figures as in example **12.1** the mass flow curve is shown in Fig. 12.2 but the slope of the demand line OJ must be adjusted so that when the reservoir is empty at B the distance BE represents the initial volume stored. This is achieved by setting off OG equal to the initial volume stored and drawing GB tangent to the curve. Draw OJ parallel to GB. Then OJ is the demand line required and

Maximum possible uniform rate of draw-off
$$= \text{slope of OJ} = 28 \cdot 1 \times 10^6 \, \text{m}^3/\text{day}$$

(*b*) Draw FH parallel to GB so that the vertical distance between FH and GB equals the reservoir capacity of $41 \times 10^6 \, \text{m}^3$. Then at F the reservoir is full. Draw a tangent DK parallel to FH and the demand line through the crest of the curve at D. From F to D the inflow exceeds the outflow, since the slope of the curve is greater than that of the demand line and water flows to waste since the reservoir is already full.

Volume flowing to waste $= \text{FL} = 14 \times 10^6 \, \text{m}^3$

Note that the reservoir is full at D and from D to A will be emptying.

12.3 Hydrograph

> The capacity of a reservoir is $425 \times 10^6 \, \text{m}^3$ and the following numbers represent the volume entering the reservoir in successive 30-day periods the units being $10^6 \, \text{m}^3$: 345, 374, 283, 153, 125, 62, 85, 68, 187, 254, 283, 334. By drawing a hydrograph determine the power which can be developed continuously if the available head is $18 \cdot 3$ m and the overall efficiency of the system is 70 per cent.

Solution. A hydrograph is a diagram in which the *rate* of inflow is plotted against time as in Fig. 12.3 so that the area under the curve $= \text{rate of flow} \times \text{time} = \text{volume}$.

For a uniform draw-off the demand line is horizontal as at BC. From A to B the supply is greater than the demand and the reservoir is filling or surplus is going to waste. From B to C the demand is greater than the supply and the reservoir is emptying. The shaded area represents the total volume drawn from the reservoir and therefore the minimum capacity of reservoir for a given demand.

In the present problem the reservoir capacity is given as $425 \times 10^6 \, \text{m}^3$. Draw BC horizontally to make the shaded area represent $425 \times 10^6 \, \text{m}^3$. Then BC represents the maximum uniform demand which can be satisfied with the given reservoir capacity.

Maximum uniform demand $Q = 189 \times 10^6 \, \text{m}^3$ per 30-day period $= 73 \, \text{m}^3/\text{s}$.

If $\eta = $ overall efficiency, $w = $ specific weight of water and $H = $ available head

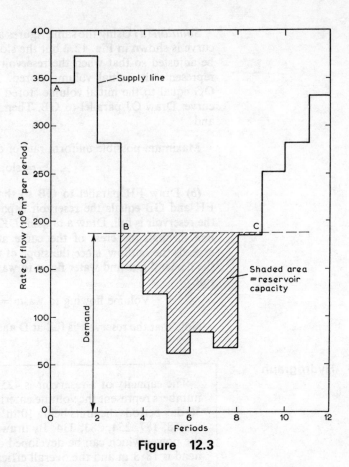

Figure 12.3

$$\text{Output power} = \eta wQH$$

$$= 0.7 \times 10^3 \times 9.81 \times 73 \times 18.3 = 9140 \times 10^3\,\text{W}$$

$$= \mathbf{9140\,kW}$$

12.4 Tabulation

A turbine installation developing 7500 kW under 27·5 m head with an overall efficiency of 83 per cent is to be supplied from a reservoir. The estimated run-off in cubic metres per month for 12 consecutive months (each of 30 days) is given by the following numbers each multiplied by 10^6: 96·2, 101·8, 86·3, 74·9, 67·9, 80·6, 113·2, 90·5, 86·3, 93·4, 99·0, 89·1. Assuming that the reservoir is full at the beginning of the first month, determine its minimum capacity to ensure the required demand. Find also the total quantity of water wasted during the year.

Solution. If Q = demand in m³/s, H the available head and η = the efficiency

$$\text{Brake power} = \rho g Q H \eta$$

$$\text{Demand in m}^3\text{/month} = Q = \frac{\text{Power}}{\rho g H \eta} = \frac{7500 \times 10^3}{10^3 \times 9 \cdot 81 \times 27 \cdot 5 \times 0 \cdot 83}$$

$$= 33 \cdot 5 \, \text{m}^3\text{/s} = 86 \cdot 8 \times 10^6 \, \text{m}^3\text{/month}$$

As the demand and monthly supply are known and the reservoir is full at the start the problem can be solved by a simple table (Table 12.2). Since the reservoir is full at the start the surplus of supply over demand in months 1 and 2 goes to waste. In month 3 demand exceeds supply and $0 \cdot 5 \times 10^6 \, \text{m}^3$ are drawn from the reservoir and there is no wastage. By the end of month 6 a total of $37 \cdot 5 \times 10^6 \, \text{m}^3$ have been drawn from the reservoir, thereafter supply again exceeds demand and the reservoir refills until at the end of month 11 the reservoir is full and $10 \cdot 8 \, \text{m}^3$ go to waste. From Table 12.2

$$\text{Minimum reservoir capacity} = \text{maximum deficit}$$

$$= 37 \cdot 5 \times 10^6 \, \text{m}^3$$

$$\text{Total wastage} = 37 \cdot 6 \times 10^6 \, \text{m}^3$$

Table 12.2

Monthly run-off ($10^6\,\text{m}^3$)	Demand ($10^6\,\text{m}^3$)	Month's surplus ($10^6\,\text{m}^3$)	Flow to reservoir ($10^6\,\text{m}^3$)	Reservoir deficit ($10^6\,\text{m}^3$)	Month's wastage ($10^6\,\text{m}^3$)	Total wastage ($10^6\,\text{m}^3$)
96·2	86·8	9·4	0	0	9·4	9·4
101·8	86·8	15·0	0	0	15·0	24·4
86·3	86·8	−0·5	−0·5	0·5	0	24·4
74·9	86·8	−11·9	−11·9	12·4	0	24·4
67·9	86·8	−18·9	−18·9	31·3	0	24·4
80·6	86·8	−6·2	−6·2	37·5	0	24·4
113·2	86·8	26·4	+26·4	11·1	0	24·4
90·5	86·8	3·7	+3·7	7·4	0	24·4
86·3	86·8	−0·5	−0·5	7·9	0	24·4
93·4	86·8	6·6	+6·6	1·3	0	24·4
99·0	86·8	12·2	+1·3	0	10·9	35·3
89·1	86·8	2·3	0	0	2·3	37·6

Problems

1 If in example **12.4** it is desired to increase the power output to the maximum by making use of the available water supply at a uniform rate throughout the year, find the percentage increase in reservoir capacity required and the corresponding increase in power output.

Answer 33·2 per cent, 3·6 per cent

2 The monthly flow for a river supplying a power plant in 14 successive 4-week periods of the driest year is estimated as follows: 2·77, 1·41, 1·64, 1·91, 2·27, 2·78, 1·36, 1·36, 1·23, 1·64, 3·27, 4·00, 3·51, 1·82 in millions of cubic metres. Calculate the size of reservoir to maintain the highest uniform output power throughout the year. If the head is 73·2 m and the overall efficiency is 70 per cent, calculate the power.

Assuming a reservoir only half this capacity is provided, calculate the maximum power to be supplied by an auxiliary plant to maintain the same output as before.
Answer 42·3 × 10⁶m³, 459 kW, 60·5 kW

3 Over an *initial* period of 12 months, each assumed to have the same number of days, the flow in millions of cubic metres each month from a catchment area is as follows

Jan.	Feb.	Mar.	Apr.	May	June	July	Aug.	Sept.	Oct.	Nov.	Dec.
16·04	12·54	9·90	6·74	3·37	1·10	0·28	0·31	0·48	4·10	12·59	17·66

The reservoir contains 4·53 million m³ of water at the beginning of the period.

(*a*) Determine what capacity of reservoir is required in order to maintain a maximum uniform rate of draw-off over the period, and to provide a reserve of 5·66 million m³.

(*b*) If in the *following* 12-month period the monthly flow figures are repeated, and the uniform draw-off rate is increased by 10 per cent, find how much water will flow over the spillway during the second 12-months. Assume that the reservoir reserve may be utilized.
Answer 28·5 × 10⁶m³, 18·75 × 10⁶m³

4 The estimated minimum monthly flow in a river is as follows: Jan. 7·65, Feb. 19·6, Mar. 17·3, Apr. 10·5, May 5·95, June 6·79, July 10·7, Aug. 13·0, Sept. 19·3, Oct. 17·3, Nov. 10·5, Dec. 6·79. (Unit flow = 10⁶m³.) The available fall is 42·7 m. What size of reservoir would be necessary to give the greatest continuous uniform output, and what would this power be if the plant efficiency was 70 per cent? If the largest reservoir which could be economically constructed had a capacity of 8·5 × 10⁶m³, what would then be the greatest continuous output?
Answer 13·9 × 10⁶m³, 134·5 kW, 120 kW

5 The estimated monthly run-off figures for a catchment area during its driest year are as follows, the units being 10⁶ cubic metres: Jan. 16·1, Feb. 9·35, Mar. 15·3, Apr. 16·7, May 13·3, June 8·49, July 5·95, Aug. 9·35, Sept. 4·81, Oct. 13·9, Nov. 15·3, Dec. 17·3.

The available head is 91·5 m and the continuous flow required by the turbines of a hydraulic power installation is estimated to be 4·38 m³/s. If there is at no time to be less than 2·83 × 10⁶m³ of water in the supply reservoir, determine, graphically or otherwise, the total reservoir capacity required and state the times when the reservoir is (*a*) full and (*b*) at its lowest level. Assume that during this year the overflow spillway does not come into action.
Answer 20·23 × 10⁶m³, May, Sept.

6 Find the capacity of a reservoir which must be provided to give the maximum continuous output throughout the year and determine the output if the available head is 130 m and the overall efficiency is 78 per cent, if the supply per 30 days is: 1·38, 1·71, 1·31, 0·93, 0·63, 0·58, 0·43, 0·81, 1·19, 1·28, 1·30, 1·33 in units of $10^6 m^3$. Calculate the reduction in power if a reservoir of only half maximum capacity can be provided.

Answer $1·99 \times 10^6 m^3$, 411·8 kW, 330·2 kW

13

Impulse and reaction turbines

Hydraulic motors fall into two categories: (*a*) positive displacement piston and cylinder machines which are not suitable for handling large quantities of fluid but are important in hydraulic control systems and (*b*) turbines or rotodynamic machines with which the present chapter is concerned.

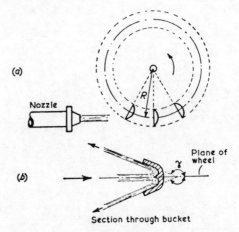

Figure   13.1

The common factor in all rotodynamic machines is that the fluid is fed to the runner or rotating element continuously in such a way that it has a tangential velocity component (or velocity of whirl) about the axis of the shaft as its enters the runner and emerges radially or axially having lost its tangential momentum and exerted a torque on the runner in the process.

In the *impulse turbine*, such as the Pelton wheel (Fig. 13.1), the energy of the fluid supplied to the machine is converted by one or more nozzles into kinetic energy. The jet strikes a series of buckets on the circumference of the wheel and is turned through an angle γ (usually 165°) thus producing a force on the bucket and a torque on the wheel. The interior of the casing of the Pelton wheel is at atmospheric pressure and is not filled with water. The wheel must be placed above tailwater level so that the water leaving the buckets falls clear of the wheel.

In the *reaction* or *pressure turbine* the fluid is fed to the runner all round the circumference from a volute casing through a ring of stationary guide vanes (Fig. 13.2), which produce a velocity of whirl. The

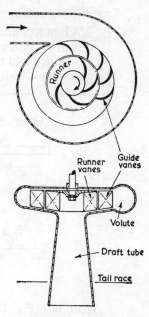

Runner vanes
Guide vanes
Volute
Draft tube
Tail race

Figure 13.2

fluid in the runner is still under pressure which is converted into kinetic
energy in the runner passages producing a reaction on the runner.
Because the water in the runner is under pressure a reaction turbine
must always run full. It need not be submerged but can be fitted
with a draft tube as shown. Since at tailrace level the pressure is
atmospheric the pressure at the exit from the runner will be below
atmospheric.

The turbine shown in Fig. 13.2 is an inward radial flow machine
known as a Francis turbine. For outward radial flow the guide vanes
are inside the runner.

Care must be taken in interpreting efficiencies as these may be
either for the turbine alone or for the whole installation including
pipeline.

$$\text{Hydraulic efficiency} = \frac{\text{work done on runner per unit weight of flow}}{\text{available head}}$$

$$\text{Overall efficiency} = \frac{\text{work delivered to shaft per unit weight of flow}}{\text{available head}}$$

$$\text{Mechanical efficiency} = \frac{\text{work delivered to shaft per unit weight of flow}}{\text{work done on runner per unit weight of flow}}$$

The shaft power is less than that done on the runner because of losses
in bearing friction and disc friction of the runner.

13.1 Pelton wheel

A Pelton wheel is supplied with water under a head of 30 m at a rate of 41 m³/min. The buckets deflect the jet through an angle of 160 deg and the mean bucket speed is 12 m/s. Calculate the power and the hydraulic efficiency of the machine.

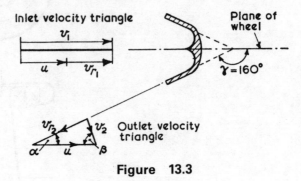

Figure 13.3

Solution. Fig. 13.3 shows the inlet and outlet velocity triangles. If H is head at the nozzle

v_1 = absolute velocity of jet at entry to the bucket

$$= \sqrt{(2gH)} = \sqrt{(2g \times 30)} = 24 \cdot 3 \, \text{m/s}$$

assuming a nozzle coefficient of discharge of unity.

u = mean bucket speed

v_{r1} = velocity of jet relative to bucket at entry

$= v_1 - u$ from the inlet triangle

In the outlet velocity triangle

v_2 = absolute velocity of the water leaving the bucket

u = mean bucket speed

v_{r2} = relative velocity of water leaving bucket

If Q = volume of water deflected per second

Force exerted on bucket

= rate of change of momentum of water in the plane of the wheel

= mass/sec deflected × change of velocity in direction of
motion of bucket

Initial absolute velocity of water in direction of motion of bucket = v_1
Component of final absolute velocity in this direction = $v_2 \cos \beta$
Change of absolute velocity in this direction = $v_1 - v_2 \cos \beta$

$$\text{Force exerted on bucket} = \rho Q(v_1 - v_2 \cos \beta) \qquad (1)$$

From the outlet velocity triangle

$$v_2 \cos \beta = u - v_{r2} \cos \alpha = u - v_{r2} \cos (180 - \gamma)$$

where γ is the deflection angle.

If there is no friction on the surface of the bucket the water enters and leaves with the same relative velocity so that

$$v_{r2} = v_{r1} = v_1 - u$$

and
$$v_2 \cos \beta = u - (v_1 - u) \cos (180 - \gamma)$$

Force exerted on bucket

$$= \rho Q[v_1 - \{u - (v_1 - u) \cos (180 - \gamma)\}]$$
$$= \rho Q(v_1 - u)\{1 + \cos (180 - \gamma)\}$$

Power = work done/sec = force on bucket × bucket speed
$$= \rho Q u(v_1 - u)\{1 + \cos (180 - \gamma)\}$$

Putting $Q = 41/60\,\text{m}^3/\text{s}$

$$v_1 = \sqrt{(2g \times 30)} = 24 \cdot 3\,\text{m/s}, \quad u = 12\,\text{m/s}, \quad \gamma = 160°$$

$$\text{Power} = 1000 \times \frac{41}{60} \times 12 \times 12 \cdot 3\,(1 + 0 \cdot 937)\,\text{W}$$

$$= \mathbf{195 \cdot 5\,kW}$$

Power supplied to the nozzle = weight/sec × head at nozzle

$$= \rho g Q H$$

Hydraulic efficiency $= \dfrac{\text{power output}}{\text{power supplied}}$

$$= \frac{\rho Q u(v_1 - u)\{1 + \cos (180 - \gamma)\}}{\rho g Q H}$$

$$= \frac{u}{gH}(v_1 - u)\{1 + \cos (180 - \gamma)\}$$

$$= \frac{12}{9 \cdot 81 \times 30} \times 12 \cdot 3 \times 1 \cdot 937$$

$$= 0 \cdot 97 \text{ or } \mathbf{97 \text{ per cent}}$$

13.2 Pelton wheel

In the theory of a Pelton wheel the following assumptions may be made: (i) the coefficient of velocity C_v of the nozzle is constant; (ii) the output power is a constant fraction ε of the power imparted by the water; and (iii) that the relative velocity of the water to the bucket at exit is n times the relative velocity at inlet, where n is a constant.

Working from these assumptions, if γ is the deflecting angle of the bucket and k denotes the ratio of bucket velocity to jet velocity, show that the graph of efficiency against k is a parabola and find the law in terms of the constants given. Show also that the maximum efficiency occurs when $k = 0\cdot5$.

Using the foregoing theory, find the output power from a Pelton wheel when the flow is reduced by 20 per cent by means of a throttle valve before the nozzle, if the wheel developed 410 kW at maximum efficiency before the flow reduction. The speed of the wheel is the same before and after the reduction of flow and the nozzle opening remains unchanged.

Solution. The working is similar to example **13.1** from which by equation (1)

$$\text{Force exerted on bucket} = \rho Q(v_1 - v_2 \cos \beta)$$

From the outlet velocity triangle (Fig. 13.3)

$$v_2 \cos \beta = u - v_{r2} \cos (180 - \gamma)$$

and in the present problem $v_{r2} = nv_{r1} = n(v_1 - u)$

Thus $\text{Force on bucket} = \rho Q(v_1 - u)\{1 + n \cos (180 - \gamma)\}$

$$\text{Work done by water/sec} = \rho Q u(v_1 - u)\{1 + n \cos (180 - \gamma)\}$$

If H = head available, $v_1 = C_v \sqrt{(2gH)}$

$$\text{Energy supplied/sec} = \rho g Q H = \rho g Q \frac{v_1^2}{C_v^2 \times 2g}$$

$$\text{Efficiency} = \eta = \frac{\text{output/sec}}{\text{energy supplied/sec}}$$

$$\text{Output power} = \varepsilon \times \text{work done by water/sec}$$

$$= \varepsilon \rho Q u (v_1 - u)\{1 + n \cos (180 - \gamma)\}$$

Thus $\eta = \dfrac{2\varepsilon u C_v^2}{v_1^2} (v_1 - u)\{1 + n \cos (180 - \gamma)\}$

and putting $u/v_1 = k$

$$\eta = 2\varepsilon k C_v^2 (1 - k)\{1 + n \cos (180 - \gamma)\}$$

The graph of η against k is therefore a parabola.

For maximum efficiency $k(1 - k)$ is a maximum:

$$\frac{d}{dk}\{k(1-k)\} = 0$$

$$1 - 2k = 0$$

$$k = \mathbf{0 \cdot 5}$$

At full power. Let $V_1 =$ jet velocity.

Volume flowing/sec $= Q_1 = aV_1$ where $a =$ jet area

$$\text{Power} = P_1 = \rho aV_1 \times u(V_1 - u)\{1 + \cos(180 - \gamma)\}$$

At maximum efficiency $k = 0 \cdot 5$ so that $V_1 = 2u$.

$$P_1 = \rho a\{1 + \cos(180 - \gamma)\} \times 2u^3$$

At reduced flow. Let $V_2 =$ jet velocity.

Discharge $Q_2 = 0 \cdot 8Q_1$

$$aV_2 = 0 \cdot 8aV_1$$

$$V_2 = 0 \cdot 8V_1 = 1 \cdot 6u$$

$$\text{Power} = P_2 = \rho aV_2 \times u(V_2 - u)\{1 + \cos(180 - \gamma)\}$$

$$= \rho a\{1 + \cos(180 - \gamma)\} \times 0 \cdot 96u^3$$

$$\frac{P_2}{P_1} = \frac{0 \cdot 96u^3}{2u^3}$$

$$P_2 = 0 \cdot 48P_1 = 0 \cdot 48 \times 410 = \mathbf{196 \cdot 8\,kW}$$

13.3 Inward flow reaction turbine

An inward flow reaction turbine has a runner $0 \cdot 5$ m diam and 75 mm wide at inlet. The inner diameter is $0 \cdot 35$ m. The effective area of flow is 93 per cent of the gross area and the flow velocity is constant. The guide vane angle is 23°, inlet vane 93°, outlet vane 30°. Calculate the speed so that the water enters without shock, and the shaft output power when the effective supply head is 60 m. Assume hydraulic friction losses of 10 per cent and a mechanical efficiency of 94 per cent.

Solution. Fig. 13.4 shows the relation of guide vanes and runner vanes and the inlet and outlet triangles of relative velocities.

In the inlet triangle:

$v_1 =$ absolute velocity of water at inlet

$u_1 =$ tangential velocity of the runner

$f_1 =$ velocity of flow

$\quad =$ radial component of v_1

$w_1 =$ velocity of whirl $=$ tangential component of v_1

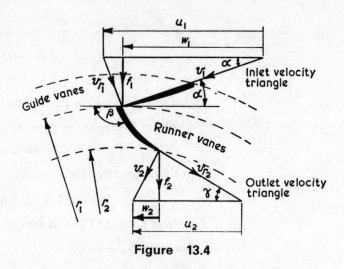

Figure 13.4

v_{r1} = velocity of water relative to runner vane

α = guide vane angle

β = inlet angle of runner vane.

In the outlet triangle:

v_2 = absolute velocity of water at exit

f_2 = velocity of flow at exit = radial component of v_2

w_2 = velocity of whirl = tangential component of v_2

u_2 = tangential velocity of runner

v_{r2} = velocity of water relative to runner vane

γ = exit angle of runner vane.

(*a*) To find the runner speed for no shock at entry. Under this condition the water enters the runner with a relative velocity parallel to the surface of runner vane. Thus v_{r1} is inclined at β to the tangent as shown.

From the inlet triangle

$$\frac{v_1}{\sin(180 - \beta)} = \frac{u_1}{\sin\{180 - \alpha - (180 - \beta)\}}$$

$$\frac{u_1}{\sin 70°} = \frac{v_1}{\sin 87°}$$

or since

$$v_1 = f_1/\sin 23°$$

$$u_1 = f_1 \frac{\sin 70°}{\sin 23° \times \sin 87°} = 2 \cdot 405 f_1$$

$$f_1 = 0 \cdot 416 u_1 \qquad (1)$$

Considering the change of momentum in the runner,
Torque on runner = rate of change of moment of momentum

If M = mass flowing per unit time,

Tangential velocity at inlet = w_1

Moment of momentum/sec at inlet

$$= \text{mass/sec} \times \text{tangential velocity} \times \text{radius}$$
$$= Mw_1r_1$$

Tangential velocity at outlet = w_2

Moment of momentum/sec at outlet = Mw_2r_2

Rate of change of moment of momentum = $M(w_1r_1 - w_2r_2)$
$$= \text{torque on runner}$$

Work done/sec on runner = torque × angular velocity
$$= M(w_1r_1 - w_2r_2)\omega$$

But $\qquad \omega r_1 = u_1 \quad \text{and} \quad \omega r_2 = u_2$

Work done/sec on runner = $M(u_1w_1 - u_2w_2)$ $\hfill$ (2)

If the effective head $H = 60\,\text{m}$ and hydraulic losses are 10 per cent,

$$0.9MgH = M(u_1w_1 - u_2w_2)$$

$$0.9 \times 60 = \frac{u_1w_1 - u_2w_2}{g} \hfill (3)$$

From equations (1) and (3), u_1 can be found by expressing w_1 and w_2 and u_2 in terms of u_1:

Angular velocity $\omega = \dfrac{u_1}{r_1} = \dfrac{u_2}{r_2}$

$$u_2 = u_1 \frac{r_2}{r_1} = u_1 \left(\frac{0.35}{0.5}\right) = 0.7u_1$$

From the inlet triangle

$$w_1 = \frac{f_1}{\tan \alpha} = 2.355f_1 = 0.98u_1$$

From the outlet triangle

$$w_2 = u_2 - \frac{f_2}{\tan \gamma} = 0.7u_1 - \frac{f_1}{\tan 30°} \quad \text{since} \quad f_1 = f_2$$
$$= 0.7u_1 - 1.73f_1 = 0.7u_1 - 0.72u_1$$
$$= -0.02u_1$$

Substituting in equation (3) and putting $H = 60\,\text{m}$

$$0.9 \times 60 = \frac{0.98u_1^2 + 0.02 \times 0.7u_1^2}{g} = \frac{u_1^2}{g}$$
$$u_1 = \sqrt{(0.9 \times 60 \times g)} = 23\,\text{m/s}$$

Speed in rev/min for no shock at inlet

$$= \frac{60u_1}{\pi d_1}$$

$$= \frac{60 \times 23}{\pi \times 0.5} = 878\,\text{rev/min}$$

(b) To find the shaft output power.

Work done per sec on runner $= M(u_1w_1 - u_2w_2) = Mu_1^2$

Mass flowing per sec $= M =$ mass density $\times$ inlet area $\times f_1$

Inlet area allowing for vane thickness

$$= \frac{93}{100} \times \pi \times 0.5 \times 0.075 = 0.11\,\text{m}^2$$

$$f_1 = 0.416u_1 = 0.416 \times 23 = 9.58\,\text{m/s}$$

Thus $\qquad M = 1000 \times 0.11 \times 9.58 = 1055\,\text{kg/s}$

Work done on runner/sec $= Mu_1^2 = 1055 \times 23^2\,\text{W}$

$$= 558\,\text{kW}$$

Mechanical efficiency $= 94$ per cent

Output shaft power $= \dfrac{94}{100} \times 558 = \mathbf{524\,kW}$

13.4 Efficiency and vane angles

Water leaves the guide vanes of an inward radial flow turbine at an angle α to the tangent to the wheel. The vane angle at entry to the wheel is 90 deg and the velocity of flow at exit is k times that at entry. Prove that for *maximum efficiency* under a head H the peripheral speed should be

$$\sqrt{\frac{2gH}{2 + k^2\tan^2\alpha}}$$

In such a turbine, rotating at 75 rev/min, the outer radius of the wheel is 0.6 m and the inner radius 0.3 m, velocity of flow at entry $= 1.8$ m/s and $k = 1$. The water is discharged radially. Calculate the vane angles at discharge and the output power developed when using 1.42 m³/s. Neglect friction in the runner.

Solution. Fig. 13.5 shows the arrangement and velocity diagrams. For conditions of maximum efficiency the flow leaves the runner radially since if H is the total energy/unit wt and p_2 the pressure at exit

$$H = \frac{p_2}{w} + \frac{v_2^2}{2g} + \text{work done/unit wt in runner}$$

The work done will be a maximum when v_2 is a minimum which from Fig. 13.4 will occur when $v_2 = f_2$ and is radial. From the inlet triangle, since $\beta = 90°$

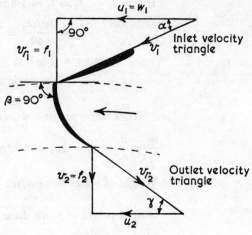

Figure 13.5

$$u_1 = w_1 \quad \text{and} \quad f_1 = u_1 \tan \alpha$$

From the outlet triangle

Velocity of whirl $w_2 = 0$ for radial exit

Also, Velocity of flow $f_2 = kf_1 = v_2$

From example **13.3**, equation (4),

$$\text{Work per unit wt/sec on runner} = \frac{u_1 w_1 - u_2 w_2}{g}$$

Putting $w_1 = u_1$, $w_2 = 0$

$$\text{Work done per unit wt/sec on runner} = \frac{u_1^2}{g} \qquad (1)$$

$$\text{Energy rejected at outlet per unit wt/sec} = \frac{v_2^2}{2g}$$

$$= \frac{k^2 f_1^2}{2g} = \frac{k^2 u_1^2 \tan^2 \alpha}{2g}$$

Input energy per unit wt/sec $= H$

Thus $$H = \frac{u_1^2}{g} + \frac{k^2 u_1^2 \tan^2 \alpha}{2g}$$

$$\text{Peripheral speed of wheel} = u_1 = \sqrt{\frac{2gH}{2 + k^2 \tan^2 \alpha}}$$

Vane angle at discharge $= \gamma$

From exit triangle $$\tan \gamma = \frac{f_2}{u_2}$$

If $k = 1$ $f_2 = f_1 = 1.8\,\text{m/s}$

$$u_2 = \frac{2\pi N r_2}{60} = \frac{2\pi \times 75}{60} \times 0.3 = 2.36\,\text{m/s}$$

$$\tan \gamma = 1.8/2.36 = 0.764$$

Vane angle at discharge $= \gamma = \mathbf{37°23'}$

Work done per unit wt/sec on runner $= \dfrac{u_1{}^2}{g}$ from equation (1) and

$$\frac{u_1}{r_1} = \frac{u_2}{r_2} \text{ or } u_1 = 2u_2 = 4.72\,\text{m/s}.$$

Weight of water flowing/sec $= \rho g Q$

Output power $=$ work done per sec $= \rho g Q \times \dfrac{u_1{}^2}{g} = \rho Q U_1{}^2$

$$= 1000 \times 1.42 \times 4.72^2\,\text{W}$$

$$= \mathbf{31.7\,kW}$$

13.5 Vane angles and head lost in runner

A vertical shaft inward flow reaction turbine runner develops 12 500 kW and uses 12·3 m³/s of water when the net head is 115 m. The runner has a diameter of 1·5 m and rotates at 430 rev/min. Water enters the runner without shock with a velocity of flow of 9·6 m/s and passes from the runner to the draft tube without whirl with a velocity of 7·2 m/s. The difference between the sum of the pressure and potential heads at the entrance to the runner and at the entrance to the draft tube is 60 m.

Determine (a) the velocity and direction of the water entering the runner from the fixed guide blades; (b) the entry angle of the runner blades; (c) the loss of head in the runner.

Also explain briefly the function of the draft tube and state what precautions must be taken with regard to its shape.

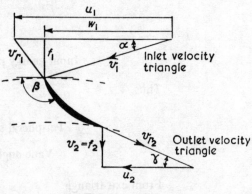

Figure   13.6

Solution. The velocity triangles are shown in Fig. 13.6.

(a) Absolute velocity of water at entry

$$v_1 = \sqrt{(f_1{}^2 + w_1{}^2)}$$

$f_1 = 9 \cdot 6\,\text{m/s}$ and w_1 is found from the power P since for radial outflow $w_2 = 0$ so that

$$P = \rho Q(u_1 w_1 - u_2 w_2) = \rho Q u_1 w_1$$

$$u_1 = \frac{\pi DN}{60} = \frac{\pi \times 1 \cdot 5 \times 430}{60} = 33 \cdot 8\,\text{m/s}$$

$$P = 12\,500 \times 10^3\,\text{W}, \quad \rho = 1000\,\text{kg/m}^3, \quad Q = 12 \cdot 3\,\text{m}^3/\text{s}$$

Thus
$$w_1 = \frac{P}{u_1 \rho Q} = \frac{12\,500 \times 10^3}{33 \cdot 8 \times 1000 \times 12 \cdot 3} = 30 \cdot 1\,\text{m/s}$$

Absolute velocity at entry $= v_1 = \sqrt{(9 \cdot 6^2 + 30 \cdot 1^2)} = \mathbf{31 \cdot 5\,m/s}$

Guide vane angle $= \alpha$

and
$$\tan \alpha = \frac{f_1}{w_1} = \frac{9 \cdot 6}{30 \cdot 1} = 0 \cdot 319 = \mathbf{17°42'}$$

(b) Entry angle of runner blades $= \beta$ and for no shock

$$\tan(180 - \beta) = \frac{f_1}{u_1 - w_1} = \frac{9 \cdot 6}{33 \cdot 8 - 30 \cdot 1} = \frac{9 \cdot 6}{3 \cdot 7} = 2 \cdot 59$$

$$180 - \beta = 68°54'$$

Entry angle $\beta = \mathbf{111°6'}$

(c) Loss of head in runner = difference in total head across runner
— energy/unit wt converted into power

If p_1 and p_2 are the pressures at inlet and outlet and z_1 and z_2 the heights of inlet and outlet above datum,

$$\text{Total energy unit wt at inlet} = \frac{p_1}{w} + z_1 + \frac{v_1{}^2}{2g}$$

$$\text{Total energy unit wt at outlet} = \frac{p_2}{w} + z_2 + \frac{v_2{}^2}{2g}$$

$$\text{Energy/unit wt converted into power} = \frac{u_1 w_1}{g}$$

$$\text{Loss of head in runner} = \frac{p_1 - p_2}{w} + z_1 - z_2 + \frac{v_1{}^2 - v_2{}^2}{2g} - \frac{u_1 w_1}{g}$$

$$\frac{p_1 - p_2}{w} + z_1 - z_2 = 60\,\text{m}, \quad v_1 = 31 \cdot 5\,\text{m/s}, \quad v_2 = f_2 = 7 \cdot 2\,\text{m/s}$$

$$\text{Loss of head in runner} = 60 + \frac{31 \cdot 5^2 - 7 \cdot 2^2}{2g} - \frac{33 \cdot 8 \times 30 \cdot 1}{g}$$

$$= 60 + 47 \cdot 8 - 103 \cdot 6 = \mathbf{4 \cdot 2\,m}$$

Figure 13.2 shows a turbine fitted with a draft tube. The purpose of the tube is to allow the turbine to be placed above tailwater level without loss of effective head and to recover some of the kinetic energy of the flow leaving the turbine for which reason it should be conical to reduce the discharge velocity to the tailrace. If the angle of flare is great the eddy losses in the diverging flow will be larger although the loss of kinetic energy at the outlet will be lower than for a smaller angle. Care is therefore required in choosing the angle which will make the combined loss a minimum.

Although at maximum efficiency the discharge should be radial at all other speeds the discharge will have a velocity of whirl and a free vortex forms in the draft tube. The low pressure at the centre of the vortex may result in air being liberated under the runner thus causing trouble. Fig. 13.7 shows a conical draft tube with a solid central core intended to prevent this.

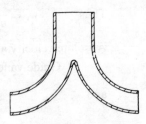

Figure   13.7

The length of the draft tube should not be excessive, (a) because the pressure at the turbine will fall further below atmospheric pressure as the length increases thus increasing the likelihood of cavitation and air release, (b) to avoid separation and waterhammer due to the inertia of the suction column if the flow is suddenly altered.

13.6 Specific speed

Derive an expression for the specific speed of a turbine in terms of its speed N, output power P and head H.

At a new hydro-electric station the available head is to be 60 m and it is anticipated that 32·3 m³/s of water will be available. Francis turbines of a specific speed of 190 are to be installed and are to run at 500 rev/min with an overall efficiency of 82 per cent. Determine the maximum power available from the turbines and the number required.

Solution. For the purpose of comparing turbines of different types and classifying them the quantity known as the *specific speed* is used and is defined as the speed in rev/min at which a turbine would operate if scaled down in geometrical proportion to such a size that it would develop 1 kilowatt under 1 metre working head.

For any turbine runner of outer diameter D and vane width B operating under a head H

Velocity of flow $f \propto \sqrt{(2gH)} \propto H^{1/2}$

Area of flow $A \propto B \times D$

or since B/D is constant for geometrically similar runners

$$A \propto D^2$$

Discharge $Q = Af \propto D^2 H^{1/2}$

Power $P \propto QH \propto D^2 H^{3/2}$

so that
$$D \propto \frac{P^{1\,2}}{H^{3/4}} \tag{1}$$

The runner speed N in rev/min will depend on the velocity of flow and on the runner diameter

$$N = \frac{\text{peripheral velocity}}{\text{runner diameter}} \propto \frac{\sqrt{H}}{D}$$

Eliminating D by using relation (1)

$$N \propto \frac{H^{5/4}}{P^{1/2}}$$

In the metric system the specific speed is defined as the value of N when $H = 1$ metre and $P = 1\,\text{kW}$. Then

$$N = N_s \frac{H^{5/4}}{P^{1/2}}$$

or   Specific speed $N_s = \dfrac{NP^{1/2}}{H^{5/4}}$

An alternative quantity is the *dimensionless specific speed* or *type number* (see Example 13.8).

Power available from turbines $= \eta Q \rho g H$

where $\eta = $ overall efficiency $= 0.82$

$\rho = $ mass density $= 1000\,\text{kg/m}^3$

$Q = 32.3\,\text{m}^3\text{/s}, \quad H = 60\,\text{m}$

Power available from turbines $= 0.82 \times 32.3 \times 10^3 \times 9.81 \times 60\,\text{W}$

$$= \mathbf{15\,650\,kW}$$

Since $N_s = NP^{1/2}/H^{5/4}$,

Power output of one turbine $= P = \left(\dfrac{N_s}{N}\right)^2 H^{5/2}$

$$= \left(\frac{190}{500}\right)^2 \times 60^{2.5} = 4020\,\text{kW}$$

Number of turbines required $= \dfrac{15\,650}{4020} = 4$

13.7 Unit speed and unit power

> Define the terms "unit speed" and "unit power" used in connexion with the operation of hydraulic turbines.
>
> Denoting the former by N_1 and the latter by P_1, show that the product $N_1\sqrt{P_1}$, termed the specific speed, is constant for all geometrically similar turbines working under dynamically similar conditions.
>
> Explain with the help of diagrams how the characteristic type and shape of turbine runner change as the value of the specific speed increases and state briefly why this change of runner shape is necessary.

Solution. Unit Speed is the theoretical speed at which a given turbine would operate under a head of 1 metre.

As shown in example **13.6**, $N \propto \dfrac{\sqrt{H}}{D}$

so that for a given turbine $\quad N \propto \sqrt{H}$

$$\text{Unit speed} = N_1 = \frac{\sqrt{H}}{N} \tag{1}$$

Unit power is the power which a given turbine would develop theoretically under a head of 1 metre.

As shown in example **13.6**, $P \propto D^2 H^{3/2}$ so that for a given turbine $P \propto H^{3/2}$.

$$\text{Unit power} = P_1 = \frac{P}{H^{3/2}}$$

For any two geometrically similar turbines working under dynamically similar conditions

$$\text{Speed } N \propto \frac{\sqrt{H}}{D} \quad \text{and} \quad \text{power } P \propto D^2 H^{3/2}$$

$$\text{Specific speed } N_s = N_1\sqrt{P_1} = \frac{N\sqrt{P}}{H^{5/4}}$$

Denoting the two turbines by suffixes A and B

$$\frac{N_{sA}}{N_{sB}} = \frac{N_A}{N_B} \times \frac{\sqrt{P_A}}{\sqrt{P_B}} \times \frac{H_B^{5/4}}{H_A^{5/4}}$$

$$= \frac{H_A^{1/2}}{D_A} \cdot \frac{D_B}{H_B^{1/2}} \times \frac{D_A H_A^{3/4}}{D_B H_B^{3/4}} \times \frac{H_B^{5/4}}{H_A^{5/4}} = 1$$

Therefore, $N_{sA} = N_{sB}$ and the specific speed is constant.

The Pelton wheel with a runner of the type shown in Fig. 13.1 has a low specific speed. For a single-jet machine N_s is about 19. If the available head and working speed are fixed the power available is determined by the specific speed. The power can be increased by increasing the flow through the machine by using two, four or six jets thus increasing the specific speed to about 60. If more than six jets

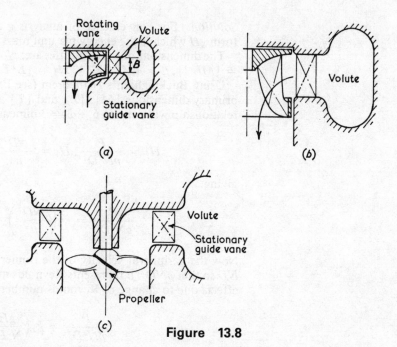

Rotating
vane

Volute

Stationary
guide vane

(a)

Volute

(b)

Volute

Stationary
guide vane

Propeller

(c)

Figure   13.8

are used they are found to interfere with each other and efficiency is reduced.

To obtain higher specific speeds and therefore higher power outputs for a given head and speed the area of flow must be increased still further. This can only be done by a complete change of design from the Pelton wheel to the Francis turbine in which the flow enters round the whole circumference of the wheel as in Fig. 13.2. This machine can be developed to have a specific speed of from 75 to 400. At low specific speed the runner width B (Fig. 13.8(a)) is small and the flow is radial. As the width B is increased to increase the flow and specific speed, purely radial flow results in difficulty at the discharge from the runner. It is found preferable to introduce the water radially and allow it to turn in the runner and discharge axially as in Fig. 13.8(b) which shows a mixed-flow turbine runner. For still greater values of specific speed, the propeller-type runner shown in Fig. 13.8(c) is used. Whirl is put into the flow by stationary vanes and the water then passes through the runner axially. An alternative arrangement has the inlet pipe axial with the shaft and a stationary set of guide vanes upstream to introduce the whirl.

13.8 Dimensionless specific speed

Derive by dimensional analysis an expression for the *dimensionless specific speed* or *type number* of a turbine in terms of the rotational speed N, power transferred between fluid and runner P, mass density ρ and viscosity μ of the fluid, the acceleration due to gravity g and the difference of head across the machine H.

Solution For convenience of analysis g and H can be combined to form gH which is the energy per unit mass of the fluid.

The dimensions of the quantities are: $N = [T^{-1}]$, $P = [ML^2T^{-3}]$, $\rho = [ML^{-3}]$, $\mu = [ML^{-1}T^{-1}]$, $gH = [L^2T^{-2}]$, $D = [L]$.

Using Buckingham's Π theorem (see Example 1.7) there are three primary dimensions $[M]$, $[L]$ and $[T]$ and six variables so that the relationship will contain $6 - 3 = 3$ dimensionless groups

$$\Pi_1 = \frac{P}{\rho N^3 D^5}, \Pi_2 = \frac{\rho N D^2}{\mu}, \Pi_3 = \frac{gH}{N^2 D^2}$$

giving

$$\frac{P}{\rho N^3 D^5} = k_1 \left(\frac{\rho N D^2}{\mu}\right)^a \left(\frac{gH}{N^2 D^2}\right)^b$$

Now the peripheral velocity of the runner is directly proportional to ND so that $\rho N D^2/\mu$ represents Reynolds number. If it is assumed that effects due to change of Reynolds number can be ignored

$$\frac{P}{\rho N^3 D^5} = k_2 \left(\frac{gH}{N^2 D^2}\right)^b$$

For a homologous series of geometrically similar machines this relationship will be independent of D, which requires that $b = 5/2$ so that

$$\frac{P}{\rho N^3 D^5} = k_2 \left(\frac{gH}{N^2 D^2}\right)^{5/2}$$

$$k_2 = \frac{(P/\rho)}{N^3 D^5} \times \frac{N^5 D^5}{(gH)^{5/2}} = \frac{N^2 (P/\rho)}{(gH)^{5/2}}$$

Taking the square root

Dimensionless specific speed $n_s = N \dfrac{(P/\rho)^{1/2}}{(gH)^{5/4}}$

13.9 Pelton wheel and pipeline

Water under a head of 270 m is available for a hydro-electric power station, and is to be delivered to the power house through three pipes $2 \cdot 4$ km long, in which the friction loss is to be 24 m.

It is decided to install a number of single-jet Pelton wheels with a specific speed not exceeding 38, to produce a total shaft output of 18 000 kW. The wheel speed is to be 650 rev/min and the ratio of bucket to jet speed $0 \cdot 46$. Assume that the overall efficiency of the wheels will be 87 per cent, and that the nozzles will have a C_d of $0 \cdot 94$ and a C_v of $0 \cdot 97$.

Determine (*a*) the number of Pelton wheels to be used, (*b*) the wheel diameter, (*c*) the jet diameter, (*d*) the diameter of the supply pipes. Assume that the friction coefficient f is $0 \cdot 006$.

Solution. Gross head $= h = 270\,\text{m}$

Head lost in friction $= h_f = 24\,\text{m}$

Power head $= H = h - h_f = 246\,\text{m}$

(a) Number of machines $= n = \dfrac{\text{total output power}}{\text{power of one machine } P} = \dfrac{18\,000}{P}$

Specific speed $N_s = \dfrac{NP^{1/2}}{H^{5/4}}$, where $N =$ wheel speed in rev/min

Assuming a specific speed of 38 and putting $N = 650\,\text{rev/min}$ and $H = 246\,\text{m}$

$$P = \left(\frac{N_s}{N}\right)^2 H^{5/2} = \left(\frac{38}{650}\right)^2 \times 246^{5/2}$$

$$= 3245 = \frac{18\,000}{n}\,\text{kW}$$

$$n = \frac{180\,000}{3245} = 5{\cdot}56, \text{ say } \textbf{6 machines}$$

(b) Jet speed $v = C_v \sqrt{(2gH)}$

$$= 0{\cdot}97\sqrt{(2g \times 246)} = 67{\cdot}1\,\text{m/s}$$

Bucket speed $= 0{\cdot}46v = 0{\cdot}46 \times 67{\cdot}1 = 30{\cdot}9\,\text{m/s}$

Also, Bucket speed $= \dfrac{\pi DN}{60} = \dfrac{\pi D \times 650}{60}$

$$\frac{\pi D \times 650}{60} = 30{\cdot}9$$

Wheel diameter $= D = \textbf{0·907\,m}$

(c) The jet diameter depends on the discharge Q per machine. For an overall efficiency of 87 per cent

Total hydraulic power required at nozzles $= \dfrac{18\,000}{0{\cdot}87} = 20\,700\,\text{kW}$

Power at nozzle per machine $= \dfrac{20\,700}{6} = 3450\,\text{kW}$

$$= 3450 \times 10^3\,\text{W} = \rho g Q H$$

$$Q = \frac{3450 \times 10^3}{1000 \times 9{\cdot}81 \times 246} = 1{\cdot}43\,\text{m}^3/\text{s}$$

If $d_j =$ jet diameter

Jet area $= \tfrac{1}{4}\pi d_j{}^2$

$$Q = C_d \tfrac{1}{4}\pi d_j{}^2 \sqrt{(2gH)}$$

$$d_j = \sqrt{\frac{4Q}{\pi C_d \sqrt{(2gH)}}} = \sqrt{\frac{4 \times 1{\cdot}43}{\pi \times 0{\cdot}94\sqrt{(2g \times 246)}}}\,\text{m}$$

Jet diameter $= d_j = \textbf{0·167\,m}$

(d) If d_p = pipe diameter, L = pipe length, V = pipe velocity

Total discharge for 6 machines = $6Q = 6 \times 1 \cdot 43 = 8 \cdot 58\,\text{m}^3/\text{s}$

Number of pipes = 3

Discharge per pipe = $Q_p = \dfrac{8 \cdot 58}{3} = 2 \cdot 86\,\text{m}^3/\text{s}$

Loss of head $h_f = \dfrac{fLQ_p^2}{3d_p^5}$

$$d_p^5 = \frac{fLQ_p^2}{3h_f} = \frac{0 \cdot 006 \times 2 \cdot 4 \times 10^3 \times 2 \cdot 86^2}{3 \times 24} = 1 \cdot 636$$

Pipe diameter = $d_p = 1 \cdot 104\,\text{m}$

Problems

1 The spouting velocity of the jet driving a Pelton wheel is 57 m/s. The wheel has a diameter of 1 m rotating at N rev/min. The relative velocity at exit from the buckets is $0 \cdot 85$ times that at entry and this relative velocity is deflected by the buckets through 165 deg. Deduce from first principles an expression for the hydraulic efficiency of the wheel. Plot a curve showing the variation of hydraulic efficiency with speed as N varies from 300 to 1000 rev/min.
Answer $3 \cdot 54\,N(1090 - N) \times 10^{-6}$

2 In a Pelton wheel the diameter of the bucket circle is $0 \cdot 9$ m and the deflecting angle of the buckets 160 deg. The jet is 75 mm diam. Neglecting friction find the power developed by the wheel and the hydraulic efficiency when the speed is 300 rev/min and the pressure behind the nozzle is 690 kN/m².
Answer 103 kW, $91 \cdot 2$ per cent

3 The bucket circle of a Pelton wheel is $1 \cdot 8$ m in diam and the deflecting angle of the buckets is 160 deg. The jet is 100 mm and the head over the nozzle is 135 m. Find the hydraulic efficiency when the speed is 250 rev/min.
Answer $96 \cdot 3$ per cent

4 A single-jet Pelton wheel with a head over the nozzle of 210 m has its buckets on a circle of $0 \cdot 9$ m diam. The deflecting angle of the buckets is 162 deg. Find (*a*) the best speed, (*b*) the hydraulic efficiency of the runner when the speed is 800 rev/min. C_v for the nozzle is $0 \cdot 975$.
Answer 664 rev/min, $93 \cdot 6$ per cent

5 A Pelton wheel driven by two similar jets transmits 3750 kW to the shaft when running at 375 rev/min. The head from reservoir level to nozzles is 200 m and the efficiency of power transmission through the pipeline and nozzles is 90 per cent. The centrelines of the jets are tangential to a $1 \cdot 45$ m diam circle. The relative velocity decreases by 10 per cent as the water traverses the bucket surfaces which are so shaped that they would, if stationary, deflect the jet through an angle of 165 deg. Neglecting windage losses, find: (a) the efficiency of the runner; (b) the diameter of each jet.

Answer 98·2 per cent, 152 mm

6 A Pelton wheel with a needle-controlled nozzle develops 710 kW when the total head is 190 m and the jet diam is 100 mm. The loss of head due to friction in the pipeline and nozzle is $92Q^2$m, where Q is the quantity of water supplied in m³/s.

Assuming that the total head and runner efficiency remain constant, determine the percentage reduction in the quantity of water supplied when the power is reduced to 630 kW by (a) alteration of the position of the needle in the nozzle, and (b) partial closure of a valve on the pipeline, the position of the needle − and, therefore, the diameter of the jet − being unchanged.

Also obtain the loss of head across the valve in case (b).

Answer 13·2 per cent, 3·9 per cent, 14·63 m

7 A double-jet Pelton wheel is supplied with water through a pipeline 1650 m long from a reservoir in which the level of the water is 375 m above that of the wheel. The turbine runs at 500 rev/min and develops 5000 kW. If the pipeline losses are 10 per cent of the gross head and $f = 0·005$, calculate the diam of the pipe, the cross-sectional area of the jets, and the mean diam of the wheel. Assume coefficient of velocity of the jets 0·98, bucket speed = 0·46 jet speed, and efficiency of turbine = 86 per cent.

Answer 0·741 m, 0·022 m², 1·4 m

8 The following data were obtained from tests on a Pelton wheel: diam of jet = 100 mm, output power = 409 kW, head at nozzle = 120 m, power loss in friction and windage = 20·9 kW, discharge = 0·39 m³/s, C_v for jet = 0·98. Assuming that the velocity of water at discharge from the buckets is 7·8 m/s, draw up a balance sheet showing the distribution of energy of the supply water.

Answer Output power 88·99 per cent, nozzle loss 3·96 per cent, windage 4·56 per cent, discharge 2·59 per cent

9 A Pelton wheel has a mean bucket speed of 12 m/s and is supplied with water at the rate of 0·68 m³/s under a head of 30 m. If the buckets deflect the jet through an angle of 160 deg find the power and efficiency of the wheel.

Answer 194 kW, 0·97

10 The following data were obtained from a test on a Pelton wheel: area of jet = 77·5 cm², head at nozzle = 30·5 m, discharge = 0·18 m³/s, output power = 41·8 kW, power absorbed in windage and friction = 2·2 kW. Determine the energy lost in the nozzle and also the energy absorbed due to losses in the wheel at discharge.

Answer 5·2 kW, 4·7 kW

11 Obtain an expression for the work done per unit wt of flow by a Pelton wheel in terms of the mean bucket velocity u, the jet velocity v and the outlet bucket angle θ neglecting all friction losses.

If the loss due to bucket friction and shock can be expressed by $k_1(v - u)^2/2g$ and that due to bearing friction and windage by $k_2u^2/2g$ where k_1 and k_2 are constants, show that the maximum efficiency occurs when

$$\frac{u}{v} = \frac{(1 - \cos \theta) + k_1}{2(1 - \cos \theta) + k_1 + k_2}$$

A Pelton wheel runner having a bucket angle of 165 deg gave, on test, a maximum efficiency of $0 \cdot 80$, u/v being $0 \cdot 47$. Find the values of k_1 and k_2 and hence express the losses as a percentage of the jet energy.
Answer $0 \cdot 232$, $0 \cdot 517$

12 A double-jet Pelton wheel required to develop 5400 kW has a specific speed of 25 and is supplied through a pipeline 790 m long from a reservoir, the level of which is 350 m above the nozzles. Allowing for 5 per cent friction loss in the pipeline calculate: (*a*) the speed in rev/min, (*b*) diameter of jets (*c*) mean diameter of bucket circle, (*d*) diameter of supply pipe.

Assume C_v for the jets is $0 \cdot 98$, bucket speed is $0 \cdot 46$ jet speed, overall efficiency of wheel 85 per cent, and f for pipe $0 \cdot 006$.
Answer 483 rev/min, 128 mm, $1 \cdot 44$ mm, $0 \cdot 807$ m

13 An inward flow pressure turbine has a runner whose vanes are radial at inlet and inclined backward at discharge. The diameter at exit is two-thirds of that at entry and the velocity of flow is constant at $4 \cdot 5$ m/s. The guide vanes are inclined at 18 deg. Determine the correct peripheral speed of the runner and the correct discharge angle of the vanes for maximum work. If the working head is 20 m what percentage of this is rejected as kinetic energy of discharge?
Answer $13 \cdot 85$ m/s, 26 deg, $5 \cdot 2$ per cent

14 In an inward flow reaction turbine the supply head is 12 m and maximum discharge $0 \cdot 28$ m³/s. External diam = $2 \times$ internal diam. Velocity of flow is constant = $0 \cdot 15\sqrt{(2gH)}$. Speed = 300 rev/min. Runner vanes are radial at inlet. Determine (*a*) guide vane angle, (*b*) vane angle at exit for radial discharge, (*c*) width of runner at inlet and outlet. Hydraulic efficiency = $0 \cdot 8$. Vanes occupy 10 per cent of the circumference.
Answer 13° 20', 25° 20', $69 \cdot 6$ mm, $139 \cdot 2$ mm

15 A mixed flow vertical reaction wheel works under a net head of 46 m and produces a shaft output of 3700 kW at an overall efficiency of 82 per cent. The shaft speed is 280 rev/min and the hydraulic efficiency is 90 per cent. The inlet to the runner is $1 \cdot 5$ m above the tailrace level and the pressure at inlet is 250 kN/m² gauge. The corresponding figures at the runner outlet are $1 \cdot 2$ m and 14 kN/m² vacuum. There is no whirl in the draft tube and the water enters at $5 \cdot 4$ m/s and leaves at 3 m/s. The runner external diameter is $1 \cdot 55$ m and the velocity of flow is 6 m/s.

Determine (a) the blade angle at inlet to the runner, (b) the exit diameter of the draft tube, and (c) the head losses in the guide blades, runner and draft tube.

Answer $51 \cdot 11°$; $2 \cdot 06$ m; $0 \cdot 89$ m, $2 \cdot 1$ m, $1 \cdot 14$ m

16 The pressure head in the casing of a reaction turbine is 48 m and the velocity head is negligible. The guide vanes make an angle of 25 deg to the tangent at entry. The moving vanes at entry are at right angles to the direction of the fixed vanes. The outer and inner diameters of the runner are 500 mm and 300 mm respectively and the widths at entrance and exit 75 mm and 125 mm respectively. The vanes occupy 6 per cent of the periphery. The pressure head inside the runner is $-2 \cdot 7$ m. Assuming the losses in guide vanes as $1 \cdot 2$ m and in the runner as $1 \cdot 8$ m of water, calculate: (a) the runner speed for no shock loss at entry; (b) the exit angles from the runner for radial discharge; (c) the power at the runner; (d) the pressure at entry to the runner.

Answer 883 rev/min, $32\frac{1}{2}$ deg, 432 kW, 250 kN/m²

17 An inward flow turbine is required to give 150 kW under a head $H = 10 \cdot 5$ m. The overall efficiency is 78 per cent and the hydraulic efficiency is 85 per cent. The turbine is to run at 150 rev/min. The velocity of flow is to be constant at $0 \cdot 2\sqrt{(2gH)}$, the peripheral velocity at inlet is to be $0 \cdot 7\sqrt{(2gH)}$ and discharge is to be radial. Determine (a) the diameter of the runner, (b) the guide vane angle, (c) the runner vane angle at inlet, (d) the width of runner at inlet assuming vane thicknesses occupy 10 per cent of the circumference.

Answer $1 \cdot 28$ m, $18° \, 15'$, $114° \, 55'$, $0 \cdot 179$ m

18 The guide vane angles of an inward radial-flow reaction turbine make 20 deg with the tangent at entry. The moving-vane angle at entry is 120 deg. The external diameter of the runner is 450 mm and the internal diameter is 300 mm. The width at entry is $62 \cdot 5$ mm and at exit 100 mm.

Calculate the exit angle of the runner so that with no shock loss at entry the discharge is radial. If the supply head in the casing is 18 m, loss in guides and runner $1 \cdot 5$ m, and there is atmospheric pressure at exit from the runner, calculate the speed in rev/min and the hydraulic efficiency. If the mechanical efficiency is 90 per cent, calculate the output power. Neglect the vane thickness.

Answer $23°$, 613 rev/min, $91 \cdot 7$ per cent, 56 kW

19 An inward flow vertical shaft reaction turbine works with a net head of 60 m and runs at 375 rev/min. The specific speed is 150 based on the power transmitted by the runner to the shaft. The external diameter of the runner is $1 \cdot 25$ m.

Water enters the runner without shock with velocity of flow of $8 \cdot 4$ m/s, passes into the draft tube without whirl with velocity of $7 \cdot 2$ m/s and is discharged from the draft tube into the tailrace with velocity of $2 \cdot 4$ m/s.

The pressure head at entrance to the runner is $28 \cdot 8$ m above and, at entrance to the draft tube, $2 \cdot 4$ m below atmospheric pressure. The mean height of the runner entry surface is $1 \cdot 8$ m and the entrance to the draft tube is $1 \cdot 5$ m above tailrace level.

Assuming a hydraulic efficiency of 90 per cent, find: (a) the entry angle of the runner blades, (b) the loss of head in (i) the volute casing and guide blades, (ii) the runner, (iii) the draft tube; (c) the entrance diameter of the draft tube.

Answer 108° 48′; $1 \cdot 8$ m, $2 \cdot 5$ m, $1 \cdot 4$ m; $1 \cdot 212$ m

20 A vertical shaft inward flow reaction turbine operates under a net head of 90 m and the runner rotates at 500 rev/min. The diameter and width of the runner at inlet are $1 \cdot 1$ m and $0 \cdot 29$ m respectively, the blade thickness coefficient is 5 per cent and the guide blade angle at entry is 18°. The velocity of flow at entry is $0 \cdot 93 \sqrt{H}$ m/s.

The water enters the elbow-type draft tube without whirl $0 \cdot 6$ m above tailrace level. The diameter of the draft tube at inlet is $1 \cdot 1$ m and the rectangular section outlet of the draft tube is $1 \cdot 5$ m deep and $2 \cdot 4$ m wide.

If the efficiency of the draft tube is assumed to be 75 per cent, the mean velocity in the tailrace beyond any disturbance from the draft tube is $0 \cdot 6$ m/s, and the overall efficiency of the turbine is $0 \cdot 95 \times$ the hydraulic efficiency, calculate: (a) the hydraulic efficiency, (b) the runner blade angle at entry surface, (c) the power developed by the turbine, (d) the pressure head at entry to the draft tube, (e) the specific speed for the turbine runner.

Answer $88 \cdot 5$ per cent, 79° 25′, 6240 kW, $-3 \cdot 71$ m, $142 \cdot 5$

21 An inward flow reaction turbine has a $1 \cdot 27$ m diam runner and operates under a net head of 61 m. The specific speed, based on the power transmitted by the runner to the shaft is 152; the hydraulic efficiency is 90 per cent; the entry area of the runner is $1 \cdot 11$ m² and the guide blade angle at entry is 20 deg. Water enters the runner without shock and leaves the runner without whirl. The velocity of flow through the runner is constant. Obtain: (a) the runner blade angle at the entry edge and at a point on the outlet edge where the radius is $0 \cdot 45$ m; (b) the runner speed in rev/min; (c) the power transmitted by the runner to the shaft.

Answer $117 \cdot 7°$, $23 \cdot 3°$, 381 rev/min, 4614 kW

22 What is meant by the specific speed of a turbine? Deduce an expression for this.

If a turbine develops 15 000 kW at 120 rev/min under a head of 18 m, what is its specific speed?

Answer 396 rev/min

23 An installation is required to supply 30 000 kW at 120 rev/min under a head of 18 m. If the proposed turbines have specific speeds of 300, how many machines must be installed?

Answer 4

24 A turbine giving 3750 kW under 12 m head runs at a design speed of 250 rev/min. It is proposed to use the same design altered to a suitable scale for a turbine giving 2250 kW under $7 \cdot 5$ m head. Calculate (a) the scale ratio for the new machine, (b)

the design speed. Derive your formula, explaining carefully the basic assumptions.

Answer 1·1 to 1; 179 rev/min

25 Develop the formula $P/H^{3/2}$ for the unit power and unit speed respectively of a turbine. Why is the unit power plotted against unit speed for a turbine when erected on site, rather than a curve of power against speed?

The following data relate to a turbine operating at 200 rev/min with full gate opening

Head (m)	7·50	6·78	6·18	5·67	5·22	4·80
Power (kW)	266	231	201	176	153	131
Efficiency	0·811	0·831	0·844	0·848	0·850	0·841

Draw graphs of unit power and efficiency against unit speed and find how much water is required per second for maximum output under a head of 6·3 m.

Answer 4·02 m³

26 Establish the expression for water turbines

$$P = \rho D^5 N^3 \phi \left(\frac{\rho D^2 N}{\mu}, \frac{DN}{\sqrt{(gH)}}, \frac{B}{D} \right)$$

where P = power developed, B and D = breadth and diameter of runner, N = speed, H = operating head, μ and ρ = viscosity and density of working fluid.

A water turbine develops 150 kW at 300 rev/min under a head of 15 m. A similar machine is to be designed to give 750 kW under a head of 18 m but otherwise under the same conditions. Find the scale ratio and speed for this machine and hence its specific speed.

Answer 1 to 1·95, 169·5 rev/min, 125

27 Define unit power and unit speed as applied to a hydraulic turbine and develop expressions for them in terms of the actual power, speed and supply head. State carefully the assumptions made.

A turbine develops 3750 kW under a supply head of 12 m and at an overall efficiency of 82 per cent. If a new source of supply is used and the head increased to 18 m, assuming the efficiency to remain at 82 per cent, find the flow in m³/s which will be required, the power which will be obtained and the percentage increase in speed.

Answer 47·65 m³/s, 6900 kW, 22·5 per cent

28 Specify the conditions necessary for dynamical similarity of operation of hydraulic turbines and explain how the formula for "unit speed", "unit power" and "specific speed" are derived.

A turbine is to develop 7500 kW at 0·85 full gate opening under a head of 30 m. A model of this turbine 300 mm diameter when running at 900 rev/min under a net head of 10·8 m gave the following test results

Fraction of gate-opening	0·4	0·6	0·8	1·0
Output power	8·5	13·5	17·7	20

Assuming the same efficiency at corresponding gate openings for the two turbines, calculate: (a) the required diameter and speed for the full-scale turbine; (b) the full-scale output at 0·5 full gate opening.

Answer   2·84 m, 159 rev/min, 4550 kW

29 The following figures relate to a test on a water turbine working under its designed head of 8·4 m

Unit power	8·9	9·3	9·57	9·57	9·4	9·1
Unit speed	56	65	75	84	93	102
Mass flow (kg/s)	3590	3540	3470	3390	3300	3200

Plot a curve of overall efficiency against unit speed. Find the turbine speed at maximum efficiency, and hence the specific speed of the machine.

If the head is changed to 9·9 m estimate the power developed and the efficiency at a turbine speed of 250 rev/min.

Answer   270 rev/min; 285; 298 kW, 83·3 per cent

30 Deduce the expression $N_s = N\sqrt{P}/H^{5/4}$ for the specific speed of a turbine.

A turbine is to be designed to give 3750 kW under a head of 15 m at 240 rev/min. At what speed should a geometrically similar model, scale one-tenth, be run under a head of 9 m and what would be its power?

Answer   1858 rev/min, 18·1 kW

14

Centrifugal pumps

A centrifugal pump consists essentially of a runner or impeller which carries a number of backward curved vanes and rotates in a casing, Fig. 14.1(*a*). Liquid enters the pump at the centre and work is done on it as it passes centrifugally outwards so that it leaves the impeller with high velocity and increased pressure. In the casing, part of the kinetic energy of the fluid is converted into pressure energy as the flow passes to the delivery pipe. Fig. 14.1(*a*) shows a volute casing which increases in area towards the delivery thus reducing the velocity of the liquid and increasing the pressure to overcome the delivery head. This type of casing has a low efficiency as there is a large loss of energy in eddies.

Figure 14.1(*b*) shows a pump with a vortex or whirlpool chamber which is a combination of a circular chamber and a spiral volute. This type of chamber has a higher efficiency of conversion of kinetic energy to pressure energy than the volute. A higher efficiency still can be obtained by using a diffuser consisting of a ring of stationary guide vanes, Fig. 14.1(*c*), an arrangement known as a *turbine pump* since it resembles a turbine operating in reverse.

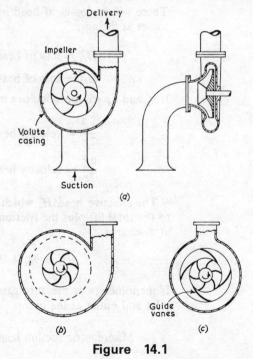

Delivery

Impeller

Volute casing

Suction

(*a*)

(*b*)

Guide vanes

(*c*)

Figure   14.1

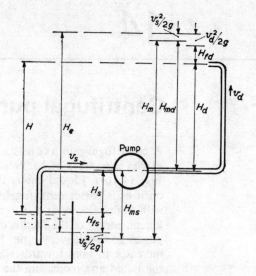

Figure 14.2

Figure 14.2 shows diagrammatically a pump with its suction and delivery pipes.

$$H_s = \text{suction lift}, \quad H_d = \text{delivery lift}$$
$$H = \text{total static head} = H_s + H_d$$

There will be losses of head in the pipelines due to friction and shock losses at fittings.

$$H_{fs} = \text{loss of head in suction pipe}$$
$$H_{fd} = \text{loss of head in delivery pipe}$$

If v_s and v_d are the velocities in the suction and delivery pipe

$$\frac{v_s^2}{2g} = \text{velocity head in suction pipe}$$

$$\frac{v_d^2}{2g} = \text{velocity head in delivery pipe}$$

The effective head H_e which the pump must provide must be equal to the total lift plus the friction loss plus the kinetic energy of the fluid at discharge:

$$H_e = H_s + H_d + H_{fs} + H_{fd} + \frac{v_d^2}{2g}$$

If manometers or pressure gauges are placed at the same level on the inlet and outlet at the pump

$$\text{Manometric suction head} = H_{ms} = H_s + H_{fs} + \frac{v_s^2}{2g}$$

$$\text{Manometric delivery head} = H_{md} = H_d + H_{fd} + \frac{v_d^2}{2g} - \frac{v_s^2}{2g}$$

$$\text{Manometric head} = H_m = H_{ms} + H_{md}$$

$$= H_s + H_d + H_{fs} + H_{fd} + \frac{v_d^2}{2g}$$

$$= \text{head rise through pump}$$

Efficiencies

Considering a pump together with suction and delivery pipes, if W is the weight discharged per sec and H is the head of fluid, then

$$\text{Overall efficiency} = \frac{\text{useful work done}}{\text{energy supplied to pump shaft}}$$

$$= \frac{WH}{\text{shaft input power}}$$

$$\text{Manometric efficiency} = \frac{\text{head rise through pump}}{\text{energy/unit weight given to fluid by impeller}}$$

$$= \frac{H_m}{u_2 w_2 / g} \quad (\textit{see} \text{ example } \mathbf{14.1})$$

$$\text{Mechanical efficiency} = \frac{\text{energy/unit wt given to fluid by impeller}}{\text{mechanical energy/unit wt supplied to shaft}}$$

14.1 Work done per unit weight and turning moment

A centrifugal pump has an impeller of outer radius r_2 and inner radius r_1 and the corresponding peripheral velocities are u_2 and u_1. If the flow enters the impeller radially obtain an expression for the work done/unit wt on the fluid by the impeller in terms of u_2 and the velocity of whirl at outlet w_2.

The diameter of the impeller of a pump is $1\cdot2$ m and its peripheral speed is 9 m/s. Water enters radially and is discharged from the impeller with a velocity whose radial component is $1\cdot5$ m/s. The vanes are curved backwards at exit and make an angle of 30 deg with the periphery. If the pump discharges $3\cdot4$ m³/min, what will be the turning moment on the shaft?

Solution Fig. 14.3 shows the triangles of velocities of the inlet and outlet of an impeller blade. To avoid shock the relative velocity at inlet should be tangential to the blade, but this will not be the case at all speeds and discharges. The relative velocity at outlet will also be tangential to the blade. The absolute velocity at outlet v_2 is found by compounding v_{r2} and u_2, but it is often convenient to consider the components of v_2 radially and tangentially which are f_2 and w_2.

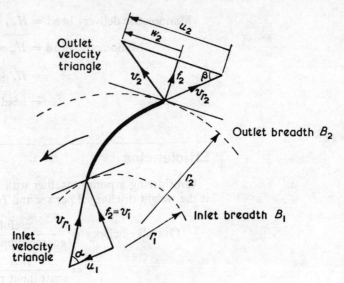

Figure 14.3

Note that the values given for blade angles in problems are often $(180° - \alpha)$ and $(180° - \beta)$.

$$\text{Discharge} = \text{peripheral area} \times \text{radial velocity}$$
$$Q = 2\pi r_1 B_1 f_1 = 2\pi r_2 B_2 f_2$$

If $\omega = $ angular velocity of impeller

$$u_1 = \omega r_1, \quad u_2 = \omega r_2$$

In passing through the impeller the tangential component of the absolute velocity of the fluid is changed and there is a change of moment of momentum:

Torque on impeller $= T = $ rate of change of moment of momentum.

Assuming that v_1 is radial, tangential velocity at inlet is zero for unit mass:

$$\text{Moment of momentum at inlet} = 0$$

Tangential component of absolute velocity at exit will be w_2.

$$\text{Moment of momentum at outlet} = w_2 r_2$$
$$\text{Change of moment of momentum} = w_2 r_2$$

Mass per sec passing $= \rho Q$ where $\rho = $ mass density
$$T = \rho Q w_2 r_2 \qquad (1)$$

Work done per sec $= $ torque $\times$ angular velocity
$$= \rho Q w_2 r_2 \omega = \rho Q w_2 u_2$$

Weight flowing per sec $= \rho g Q$

Work done per unit wt $= \dfrac{w_2 u_2}{g}$

From equation (1)

$$\text{Turning moment on shaft} = T = \rho Q w_2 r_2$$

From outlet velocity triangle

$$w_2 = u_2 - f_2 \cot \beta$$

Putting $u_2 = 9\,\text{m/s}$, $f_2 = 1 \cdot 5\,\text{m/s}$, $\beta = 30°$

$$w_2 = 9 - 1 \cdot 5\sqrt{3} = 6 \cdot 4\,\text{m/s}$$

Also $\qquad \rho = 1000\,\text{kg/m}^3, \quad Q = \dfrac{3 \cdot 4}{60}\,\text{m}^3/\text{s}, \quad r_2 = 0 \cdot 6\,\text{m}$

$$\text{Torque on shaft} = 1000 \times \frac{3 \cdot 4}{60} \times 6 \cdot 4 \times 0 \cdot 6$$

$$= \mathbf{217\,N\text{-}m}$$

14.2 Speed to commence pumping

If the static lift for a centrifugal pump is h metres, the speed of rotation N rev/min and the external diameter of the impeller d metres, deduce the expression

$$N = 83 \cdot 5 \; \frac{\sqrt{h}}{D}$$

for the speed at which pumping commences, assuming only rotation of water in the impeller at the "no-flow" condition.

Such a pump delivers $1 \cdot 27\,\text{m}^3$ of water per minute at 1200 rev/min. The impeller diameter is 350 mm and breadth at outlet $12 \cdot 7\,\text{mm}$. The pressure difference between inlet and outlet flanges is $272\,\text{kN/m}^2$. Taking the manometric efficiency at 63 per cent, calculate the impeller exit blade angle.

Solution. Under no-flow conditions a forced vortex is formed by the impeller. Pumping can commence when the pressure difference from centre to outside of vortex is equal to the static lift or

$$\frac{u_2{}^2}{2g} = h \quad \text{where} \quad u_2 = \text{peripheral velocity}$$

so that $\qquad u_2 = \sqrt{(2gh)}$

Now $\qquad u_2 = \dfrac{\pi D N}{60}$

therefore $\qquad \dfrac{\pi D N}{60} = \sqrt{(2gh)}$

$$N = \frac{60\sqrt{(2g)}}{\pi D} \sqrt{h} = 83 \cdot 5 \frac{\sqrt{h}}{D}$$

The outlet velocity diagram will be as Fig. 14.3. The exit blade angle is β and can be found from

$$\tan \beta = \frac{f_2}{u_2 - w_2}$$

$$\text{Peripheral speed} = u_2 = \frac{\pi d_2 N}{60} = \pi \times 0.35 \times \frac{1\,200}{60} = 22\,\text{m/s}$$

$$\text{Velocity of flow} = f_2 = \frac{Q}{\pi d_2 B_2} = \frac{1.27}{60\pi \times 0.35 \times 0.0127} = 1.52\,\text{m/s}$$

Since the manometric efficiency and the pressure rise through the pump are known the work done per unit wt can be found and used to calculate w_2 (*see* example **14.1**)

$$\text{Work done/unit wt} = \frac{u_2 w_2}{g} = \frac{\text{head rise through pump}}{\text{manometric efficiency}} = \frac{p/\rho g}{\eta_m}$$

$$\therefore \qquad w_2 = \frac{p}{\rho \eta_m u_2}$$

Putting $p = 272 \times 10^3\,\text{N/m}^2$, $\rho = 10^3\,\text{kg/m}^3$, $\eta_m = 0.63$, $u_2 = 22\,\text{m/s}$

$$w_2 = \frac{272 \times 10^3}{10^3 \times 0.63 \times 22} = 19.6\,\text{m/s}$$

Thus $\qquad \tan \beta = \frac{f_2}{u_2 - w_2} = \frac{1.52}{22 - 19.6} = \frac{1.52}{2.4} = 0.633$

$$\text{Exit blade angle} = \beta = 32°20'$$

14.3 Efficiency and losses

A centrifugal blower has an impeller of outer diameter 500 mm and width 75 mm with vanes set back at 70 deg to the tangent at the outer periphery. When the blower is delivering air weighing $1 \cdot 25$ kg/m³ at a rate of $3 \cdot 1$ m³/s the speed is 900 rev/min and the pressure difference across the blower measured by a manometer is 33 mm of water. The power supplied to the blower shaft is $1 \cdot 65$ kW and the mechanical efficiency is 93 per cent.

Assuming radial inlet to the impeller and neglecting the thickness of the vanes, find the manometric and the overall efficiencies. Also determine the power lost in (a) bearing friction and windage, (b) the diffuser and (c) the impeller.

Solution. Manometric efficiency $= \dfrac{\text{manometric head}}{\text{work done/unit wt in impeller}}$

Since the fluid is air, express the manometric pressure difference as a head of air.

$$\text{Manometric head} = 0.033 \times \frac{10^3}{1.25} = 26.4\,\text{m of air}$$

From example **14.1**

$$\text{Work done per unit wt in impeller} = \frac{u_2 w_2}{g}$$

The outlet velocity is similar to Fig. 14.3.

$$\text{Peripheral velocity at outlet} = u_2 = \frac{\pi d_2 N}{60}$$

$$= \pi \times 0.5 \times \frac{900}{60} = 23.55\,\text{m/s}$$

$$w_2 = u_2 - f_2 \cot \beta$$

$$f_2 = \frac{\text{discharge}}{\text{outlet area}} = \frac{Q}{\pi d_2 B_2} = \frac{3.1}{\pi \times 0.5 \times 0.075} = 26.35\,\text{m/s}$$

Thus $\quad w_2 = 23.55 - 26.35 \cot 70°$

$$= 23.55 - 9.6 = 14.05\,\text{m/s}$$

Work done/unit wt in impeller

$$= \frac{u_2 w_2}{g} = \frac{23.55 \times 14.05}{9.81} = 33.8\,\text{J/N}$$

$$\text{Manometric efficiency} = \frac{H_m}{u_2 w_2/g} = \frac{26.4}{33.8} = \textbf{78.3 per cent}$$

Weight of air delivered per sec $= W = \rho g Q = 1.25 \times 9.81 \times 3.1\,\text{N/s}$

Output power of blower $= W H_m = 38.1 \times 26.4\,\text{N-m/s} = 1.005\,\text{kW}$

Mechanical input to shaft $= 1.65\,\text{kW}$

$$\text{Overall efficiency} = \frac{\text{output}}{\text{input}} = \frac{1.005}{1.65} = \textbf{60.9 per cent}$$

The losses are:

(a) The mechanical efficiency is 93 per cent, therefore

$$\text{Bearing and windage loss} = \tfrac{7}{100} \times \text{power supplied}$$

$$= 0.07 \times 1.65 = \textbf{0.115 kW}$$

(b) The loss in the diffuser is the difference between the power put into the fluid leaving the impeller and the output power.

$$\text{Work done in impeller per sec} = W \times \frac{u_2 w_2}{g}$$

$$= 38.1 \times 33.8 = 1290\,\text{W} = 1.29\,\text{kW}$$

$$\text{Output power} = 1.005\,\text{kW}$$

$$\text{Power lost in diffuser} = 1.29 - 1.005 = \textbf{0.285 kW}$$

(c) Loss in impeller $=$ power supplied $-$ bearing loss

$$-\text{ diffuser loss} - \text{output power}$$

$$= 1.65 - 0.115 - 0.285 - 1.005$$

$$= 1.65 - 1.405 = \textbf{0.245 kW}$$

A centrifugal pump running at 700 rev/min is supplying 9 m³/min against a head of 19·8 m. The blade angle at exit is 135° from the direction of motion of the blade tip. Assume that the relative velocity of the water at exit is along the blade and that the absolute velocity at inlet is radial. The velocity of flow is constant at 1·8 m/s. Calculate the necessary impeller diameter (a) if none of the energy corresponding to the velocity at the exit from the impeller is recovered; (b) if 40 per cent of this energy is recovered.

In case (b) find also the width of impeller at exit, allowing 8 per cent for vane thickness.

Solution The velocity diagram at outlet is similar to Fig. 14.3, with $\beta = 180° - 135° = 45°$.

(a) If no kinetic energy is recovered

Work done/unit wt in impeller = static head + velocity head

$$\frac{u_2 w_2}{g} = H + \frac{v_2{}^2}{2g} \tag{1}$$

From the outlet triangle

$$w_2 = u_2 - f_2 \cot 45°, \quad \text{put} \quad f_2 = 1·8 \, \text{m/s}$$
$$w_2 = u_2 - 1·8$$

Also $\quad v_2{}^2 = f_2{}^2 + w_2{}^2 = 1·8^2 + (u_2{}^2 - 3·6u_2 + 1·8^2)$
$$= u_2{}^2 - 3·6u_2 + 6·48$$

Substituting in equation (1) and multiplying both sides by 2g,

$$2u_2(u_2 - 1·8) = 2gH + (u_2{}^2 - 3·6u_2 + 6·48)$$

Putting $H = 19·8$ m and solving for u_2

$$2u_2{}^2 - 3·6u_2 - (u_2{}^2 - 3·6u_2 + 6·48) = 2 \times 9·81 \times 19·8$$
$$u_2{}^2 = 389 + 6·48 = 395·48$$
$$u_2 = 19·9 \, \text{m/s}$$

$$\text{Impeller diameter} = \frac{u_2 \times 60}{\pi N}$$

$$= \frac{19·9 \times 60}{\pi \times 700} = \textbf{0·542 m}$$

(b) If 40 per cent of the kinetic energy is recovered, 60 per cent is lost.

Work done/unit wt in impeller = static head + 0·6 velocity head

$$\frac{u_2 w_2}{g} - 0·6 \frac{v_2{}^2}{2g} = H$$

$$2u_2(u_2 - 1·8) - 0·6(u_2{}^2 - 3·6u_2 + 6·48) = 19·8 \times 2g$$

$$1 \cdot 4 u_2{}^2 - 1 \cdot 44 u_2 - 399 \cdot 37 = 0$$

$$u_2 = 17 \cdot 35 \, \text{m/s}$$

$$\text{Impeller diameter} = \frac{u_2 \times 60}{\pi N}$$

$$= \frac{17 \cdot 35 \times 60}{\pi \times 700} = \mathbf{0 \cdot 473 \, m}$$

Effective area at outlet is reduced by 8 per cent owing to blade thickness.

$$\text{Discharge} = \text{effective peripheral area} \times \text{radial velocity}$$

$$Q = 0 \cdot 92 \pi D_2 B_2 f_2$$

$$B_2 = \frac{Q}{0 \cdot 92 \pi D_2 f_2}$$

Putting $Q = 9 \, \text{m}^3/\text{min} = 9/60 \, \text{m}^3/\text{s}$

$$D_2 = 0 \cdot 473 \, \text{m}, \quad f_2 = 1 \cdot 8 \, \text{m/s}$$

$$\text{Width at exit} = B_2 = \frac{9}{60 \times 0 \cdot 92 \times \pi \times 0 \cdot 473 \times 1 \cdot 8} \, \text{m}$$

$$= 0 \cdot 061 \, \text{m} = \mathbf{61 \, mm}$$

14.5 Specific speed

Explain what is meant by the specific speed of a centrifugal pump and show that its value is $NQ^{1/2}/H^{3/4}$ where N is the rotational speed of the impeller, Q the discharge and H the operating head.

A centrifugal pump, having four stages in parallel, delivers 11 m³/min of liquid against a head of 24·7 m, the diameter of the impellers being 225 mm and the speed 1700 rev/min.

A pump is to be made up with a number of identical stages in series, of similar construction to those in the first pump, to run at 1250 rev/min and to deliver 14·5 m³/min against a head of 248 m. Find the diameter of the impellers and the number of stages required.

Solution. The *specific speed* is used as a basis of comparison of the performance of different pumps and is defined as the theoretical speed at which the given pump would deliver unit quantity against unit head. For example the speed in rev/min at which the pump would discharge 1 m³/min under 1 metre head. The specific speed of a given pump depends on the system of units chosen.

To find this theoretical speed for unit discharge under unit head it is necessary to scale down the operating values for the pump. This is done by assuming that in scaling-down proportions are kept geometrically similar and all linear dimensions are proportional to the impeller diameter. It is also assumed that the velocity diagrams are similar and all velocities are proportional to the square root of the head H.

Breadth of impeller $B \propto$ diameter D

Impeller velocity $u \propto H^{1/2}$

Also if the speed of the impeller is N rev/min

$$u \propto ND \quad \text{or} \quad D \propto \frac{u}{N}$$

or
$$D \propto \frac{H^{1/2}}{N}$$

Discharge $Q \propto$ area of flow $\times$ velocity of flow

$\propto \pi DBf$

Now
$$f \propto H^{1/2} \quad \text{and} \quad B \propto D$$

so that
$$Q \propto D^2 H^{1/2}$$

Substituting,
$$D \propto \frac{H^{1/2}}{N}$$

$$Q \propto \frac{H}{N^2} H^{1/2}$$

$$N = N_s \frac{H^{3/4}}{Q^{1/2}}$$

Specific speed $N_s = \dfrac{NQ^{1/2}}{H^{3/4}}$

Considering pump with 4 stages in parallel

Discharge for one stage $= \dfrac{11}{4}\,\text{m}^3/\text{min}$

$$Q_1 = 2 \cdot 75\,\text{m}^3/\text{min}$$

Operating head $H_1 = 24 \cdot 7\,\text{m}$

Operating speed $N_1 = 1\,700\,\text{rev/min}$

Specific speed $= \dfrac{N_1 Q_1^{1/2}}{H_1^{3/4}} = \dfrac{1700 \times \sqrt{2 \cdot 75}}{24 \cdot 7^{3/4}} = \mathbf{254}$

For the multi-stage pump

If each stage is similar to those of the first pump

Specific speed of each stage $= N_s = 254$

The whole discharge passes through each stage, so that

$$Q_2 = 14 \cdot 5\,\text{m}^3/\text{min}$$

$$N_2 = 1\,250\,\text{rev/min}$$

$$N_s = \frac{N_2 \sqrt{Q_2}}{H_2^{3/4}}$$

where $H_2 =$ head rise per stage.

$$254 = \frac{1250\sqrt{14 \cdot 5}}{H_2^{3/4}}$$

$$H_2^{3/4} = 18 \cdot 7; \quad H_2 = 49 \cdot 64 \, \text{m}$$

$$\text{Total head required} = 248 \, \text{m}$$

$$\text{Number of stages required} = \frac{248}{H_2} = 5$$

Since the head H is proportional to the square of the impeller velocity u and $u \propto ND$,

$$H = kN^2 D^2$$

Comparing the original pump and one stage of the second pump, for similarity

$$\frac{H_1}{H_2} = \left(\frac{N_1}{N_2}\right)^2 \left(\frac{D_1}{D_2}\right)^2$$

Thus

$$D_2 = D_1 \frac{N_1}{N_2} \sqrt{\frac{H_2}{H_1}}$$

$D_1 = 0 \cdot 225 \, \text{m}, \, N_1 = 1700 \, \text{rev/min}, \, N_2 = 1250 \, \text{rev/min}.$
$H_1 = 24 \cdot 7 \, \text{m}, \, H_2 = 49 \cdot 64 \, \text{m}.$

$$\text{Diameter of impeller } D_2 = 0 \cdot 225 \times \frac{1700}{1250} \sqrt{\frac{49 \cdot 64}{24 \cdot 7}} = 0 \cdot 433 \, \text{m}$$

$$= \mathbf{433 \, mm}$$

14.6 Type number or dimensionless specific speed

(a) By dimensional analysis derive expressions for the head coefficient K_H, flow coefficient K_Q, and power coefficient K_p of a centrifugal pump or fan and explain how these can be combined to give the type number or dimensionless specific speed.

(b) A centrifugal pump running at 2950 rev/min under test at peak efficiency gave the following results: Effective head $H = 75$ m of water, Rate of flow $Q = 0 \cdot 05$ m³/s, overall efficiency $\eta = 76$ per cent. Calculate the dimensionless specific speed of this pump based on rotational speed in rev/s.

(c) A dynamically similar pump is to operate at a corresponding point of its characteristic when delivering $0 \cdot 5$ m³/s through a pipe 800 m long and 1 m diameter for which the friction coefficient $f = 0 \cdot 05$. The pipe discharges 90 m above reservoir level. Determine the rotational speed at which the pump should run to meet the duty and the ratio of its impeller diameter to that of the pump in (b), stating any assumptions made. What will be the power consumed by the pump.

Solution (a) For the general dimensional analysis of any rotodynamic machine the variables to be considered are:

$Q =$ Volumetric flow rate $[L^3 T^{-1}]$
$P =$ Power transferred from impeller to fluid $[ML^2 T^{-3}]$
$N =$ Rotational speed of the impeller $[T^{-1}]$
$H =$ Difference of head across machine $[L]$

D = Diameter of impeller [L]
ρ = Density of fluid [ML^{-3}]
μ = Dynamic viscosity of fluid [$ML^{-1}T^{-1}$]
K = Bulk modulus of elasticity of fluid [$ML^{-1}T^{-2}$]
ε = Roughness of internal passages represented by a typical dimension [L]

The head H is the energy per unit weight of the fluid and it is convenient to substitute gH which is the energy per unit mass.

Using Buckingham's theorem (see example **1.7**) there are nine variables and three fundamental dimensions therefore there will be six dimensionless groups in the relationship. These can be

$$\Pi_1 = \frac{gH}{N^2D^2} \qquad \text{known as the head coefficient } K_H$$

$$\Pi_2 = \frac{Q}{ND^3} \qquad \text{known as the flow coefficient } K_Q$$

$$\Pi_3 = \frac{P}{N^3D^5\rho} \qquad \text{known as the power coefficient } K_P$$

$$\Pi_4 = \frac{\mu}{ND^2\rho} \qquad \begin{array}{l}\text{which, since } ND \text{ is the peripheral velocity, is proportional to } 1/Re \text{ where } Re \text{ is Reynolds number based on impeller diameter.}\end{array}$$

$$\Pi_5 = \frac{K}{N^2D^2\rho} \qquad \begin{array}{l}\text{which, since } \sqrt{K/\rho} \text{ is the velocity of sound } a, \text{ is proportional to } 1/Ma \text{ where } Ma \text{ is the Mach number.}\end{array}$$

$$\Pi_6 = \frac{\varepsilon}{D} \qquad \begin{array}{l}\text{which is the relative roughness of the internal passages of the machine.}\end{array}$$

Now

$$\Pi_1 = \phi\{\Pi_2, \Pi_3, \Pi_4, \Pi_5, \Pi_6\}$$

or

$$\frac{gH}{N^2D^2} = \phi\left\{\frac{Q}{ND^3}, \frac{P}{N^3D^5\rho}, \frac{\mu}{ND^2\rho}, \frac{K}{N^2D^2\rho}, \frac{\varepsilon}{D}\right\}$$

or

$$K_H = \phi(K_Q, K_P, Re, Ma, \varepsilon/D)$$

Comparison of rotodynamic machines can be made on the basis of the values of K_Q, K_H and K_P. For pumps K_Q and K_H are the most important factors and their ratio K_Q/K_H indicates whether a particular pump is suitable for large or small flows for a given head. For geometrical similar machines the impeller diameter can be eliminated by using the ratio of $K_Q^{1/2}$ to $K_H^{3/4}$ which is known as the *type number* or *dimensionless specific speed* n_s

$$n_s = \frac{(K_Q)^{1/2}}{(K_H)^{3/4}} = \left(\frac{Q}{ND^3}\right)^{1/2}\left(\frac{N^2D^2}{gH}\right)^{3/4}$$

$$n_s = N \frac{Q^{1/2}}{(gH)^{3/4}}$$

The value of n_s which is the type number is calculated from the values of N, Q and H corresponding to the *design point*, that is to say the particular duty for which the machine is designed.

(b) Putting $N = 2950$ rev/min $= 49 \cdot 17$ rev/s

$$Q = 0 \cdot 05 \text{ m}^3/\text{s} \quad \text{and} \quad H = 75 \text{ m of water}$$

$$n_s = 49 \cdot 17 \times \frac{0 \cdot 05^{1/2}}{(9 \cdot 81 \times 75)^{3/4}} = 7 \cdot 79 \times 10^{-2}$$

(c) Lift of pump $= 90$ m

$$\text{Friction loss in pipe} = h_f = \frac{fLQ^2}{3d^5}$$

$$= \frac{0 \cdot 05 \times 800 \times (0 \cdot 45)^2}{3 \times (1)^5} = 27 \text{ m}$$

Effective head required $= H_2 = h + h_f = 90 + 27 = 117$ m

For operation at the same point of the characteristic curve n_s will be the same. Therefore $n_s = 7 \cdot 79 \times 10^{-2}$. Substituting in the expression for specific speed

$$7 \cdot 79 \times 10^{-2} = N \frac{(0 \cdot 45)^{1/2}}{(9 \cdot 81 \times 117)^{3/4}} = 3 \cdot 4 \times 10^{-3} N$$

$$N = \frac{7 \cdot 79}{0 \cdot 34} \text{ rev/s} = \textbf{1375 rev/min}$$

Since the head coefficient must be the same for both pumps

$$\frac{gH_1}{(N_1 D_1)^2} = \frac{gH_2}{(N_2 D_2)^2}$$

$$\frac{D_2}{D_1} = \left(\frac{H_2}{H_1}\right)^{1/2} \left(\frac{N_1}{N_2}\right) = \left(\frac{117}{75}\right)^{1/2} \left(\frac{2950}{1375}\right) = \textbf{2·68}$$

Assuming no scale effects and no variation in efficiency

$$\text{Power consumed by pump} = \frac{\text{power transferred to fluid}}{\text{efficiency}}$$

$$= \frac{\rho g Q H}{\eta}$$

$$= \frac{1000 \times 9 \cdot 81 \times 0 \cdot 45 \times 117}{0 \cdot 7} \text{ W}$$

$$= \textbf{736 kW}$$

14.7 Performance of pump and pipeline

A centrifugal pump running at 1000 rev/min gave the following relation between head and discharge:

Discharge (m³/min)	0	4·5	9·0	13·5	18·0	22·5
Head (m)	22·5	22·2	21·6	19·5	14·1	0

The pump is connected to a 300 mm suction and delivery pipe the total length of which is 69 m and the discharge to atmosphere is 15 m above sump level. The entrance loss is equivalent to an additional 6 m of pipe and f is assumed as 0·006. Calculate the discharge in m³ per minute.

If it is required to adjust the flow by regulating the pump speed, estimate the speed to reduce the flow to one-half.

Solution.

Head required from pump = static + friction + velocity head

$$H = 15 + H_f + H_v$$

Both H_f and H_v depend on the discharge Q so that both the head required and the head available are functions of the discharge and if these are plotted on a base of quantity the intersection of the two curves will give the discharge required.

$$\text{Head lost in friction} = H_f = \frac{fLQ^2}{3d^5}$$

$$= \frac{0{\cdot}006 \times (69 + 6)Q^2}{3 \times (0{\cdot}3)^5}$$

where Q is in m³/s
If q = discharge in m³/min, then $q = 60Q$.

$$\therefore \qquad Q^2 = \frac{q^2}{3600}$$

$$\text{Head lost in friction} = \frac{0{\cdot}006 \times 75 \times q^2}{3 \times (0{\cdot}3)^5 \times 3600} = 17{\cdot}15 \times 10^{-3}q^2$$

$$\text{Velocity head} = H_v = \frac{v^2}{2g} = \frac{Q^2}{2g\left(\frac{\pi}{4}d^2\right)^2}$$

$$= \frac{q^2 \times 16}{3600 \times 2g \times \pi^2 \times (0{\cdot}3)^4}$$

$$= 2{\cdot}83 \times 10^{-3}q^2$$

Head required = $H = 15 + 19{\cdot}98 \times 10^{-3}q^2$ where q is in m³/min

From this expression and the figures given in the problem the following table is compiled:

Table 14.1

Discharge q (m³/min)	0	4·5	9·0	13·5	18·0	22·5
Head available (m)	22·5	22·2	21·6	19·5	14·1	0
Head required (m)	15	15·4	16·6	18·6	21·5	25·1

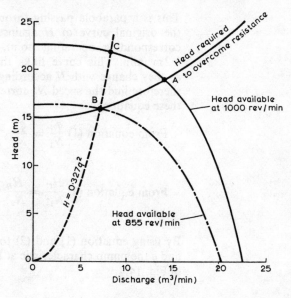

Figure 14.4

These are plotted in Fig. 14.4 from which the operating point of the system will be at the point A at which

$$\text{Head available} = \text{head required}$$
$$\text{Operating head } H_A = 19\,\text{m}, \quad \text{Discharge } q_A = \textbf{14\,m}^3/\textbf{min}$$

At reduced speed. For half-flow there will be a new operating point B at which

$$\text{Discharge} = q_B = 7\,\text{m}^3/\text{min}$$

From Fig. 14.4

$$\text{Head required to overcome resistance} = H_B = 16\cdot0\,\text{m}$$

For every speed N curves of H against q could be drawn for the pump similar to that already drawn for $N = 1000\,\text{rev/min}$. The problem therefore is to find the speed N_2 of the pump for which the corresponding curve of H against q passes through B.

Since for a given pump $q \propto N$ and $H \propto N^2$ we have

$$\frac{q}{N} = \frac{q_B}{N_2} \tag{1}$$

and

$$\frac{H}{N^2} = \frac{H_B}{N_2^2} \tag{2}$$

Eliminating N $$H = H_B \left(\frac{q}{q_B}\right)^2$$

Putting $H_B = 16.0\,\text{m}$ when $q_B = 7\,\text{m}^3/\text{min}$
$$H = \frac{16.0}{7^2} q^2 = 0.327 q^2$$

This is a parabola passing through the origin and B which intersects the original curve of H against q for $N_1 = 1000\,\text{rev/min}$ at C, the corresponding operating point, for which $H_C = 21.9\,\text{m}$ and $q_C = 8.2\,\text{m}^3/\text{min}$. This curve links the corresponding values of H and q as they change with N according to equations (1) and (2) and we can therefore find the speed N_2 corresponding to H_B and q_B from either of these equations.

From equation (1) $\dfrac{q_C}{N_1} = \dfrac{q_B}{N_2} \therefore N_2 = N_1 \dfrac{q_B}{q_C} = 1000 \times \dfrac{7}{8.2}$
$$= 855\,\text{rev/min}$$

From equation (2) $\dfrac{H_C}{N_1^2} = \dfrac{H_B}{N_2^2} \therefore N_2 = N_1 \sqrt{\dfrac{H_B}{H_C}} = 1000 \times \sqrt{\dfrac{16}{21.9}}$
$$= 855\,\text{rev/min}$$

By using equation (1) and (2) to scale down the original values of H and q the pump characteristic at 855 rev/min could be plotted as shown in Fig. 14.4.

Problems

1 The external diameter of the impeller of a centrifugal pump is 250 mm and the internal diameter is 150 mm, the corresponding width of the impeller at this latter point being 15 mm. The vane angle at outlet makes an angle of 45 deg backwards to the tangent of the impeller circle. If the radial velocity of flow is constant, the discharge being 2.7 m³/min when the speed is 1100 rev/min calculate (*a*) the impeller angle at inlet; (*b*) the angle of the guide vanes in the diffuser ring; (*c*) the pressure rise through the pump assuming a diffuser ring efficiency of 60 per cent and neglecting frictional losses.
Answer 143° 39′, 38° 27′, 11·8 m

2 A centrifugal pump when running at 1500 rev/min is to deliver 90 dm³/s against a head of 24 m. The flow at entry is radial, and the radial velocity of flow is to be constant through the impeller at 3.6 m/s. The diffuser vanes may be assumed to convert 50 per cent of the kinetic head at exit from the impeller into pressure head. The outer diameter is to be twice the inner and the width of the impeller at exit is to be 12 per cent of the diameter. Neglecting impeller losses and the influence of blade

thickness, determine the diameter and widths at inlet and outlet and the impeller and guide vane angles.
Answer 129 mm, 258 mm, 62 mm, 31 mm, 19° 33′, 28° 55′, 14° 38′

3 A centrifugal pump lifts water against a head of 36 m, the manometric efficiency being 80 per cent. The suction and delivery pipes are both of 150 mm bore, the impeller is of 375 mm diam and 25 mm wide at the outlet: its exit blade angle is 25 deg and the specified rotational speed is 1320 rev/min. If the total loss by friction in the pipeline at this speed is estimated at 9 m, calculate the probable rate of discharge at this speed.
Answer 0·06 m³/s

4 The impeller of a centrifugal pump is 325 mm diam and 19 mm wide at outlet. The blade angle at outlet is 35 deg, wheel speed 1600 rev/min, suction lift 1·5 m, and estimated loss of head on the suction side 2·1 m. The static lift from the pump centre is 39 m and the delivery pipe losses 9·6 m. If the manometric efficiency of the pump is 76 per cent and the overall efficiency 68 per cent find the discharge in dm³/s and the power needed if both the suction and delivery pipes are 125 mm diam.
Answer 35·5 dm³/s, 20·74 kW

5 A centrifugal pump impeller has an external diameter of 300 mm and discharge area of 0·11 m². The blades are bent backwards so that the direction of the relative velocity at the discharge surface makes an angle of 145 deg with the tangent to this surface drawn in the direction of impeller rotation. The diameters of the suction and delivery pipes are 300 mm and 225 mm respectively.

Gauges at points on the suction and delivery pipes close to the pump and each 1·5 m above the level in the supply sump showed heads of 3·6 m below and 18·6 m above atmospheric pressure when the pump was delivering 0·2 m³/s of water at 1200 rev/min. It required 71 kW to drive the pump.

Find (*a*) the overall efficiency; (*b*) the manometric or hydraulic efficiency, assuming that water enters the impeller without shock or whirl; (*c*) the loss of head in the suction pipe.
Answer 61·3 per cent, 71·3 per cent, 1·7 m

6 A centrifugal pump has to discharge 225 dm³/s of water and develop a head of 22·5 m when the impeller rotates at 1500 rev/min. Determine (*a*) the impeller diameter and (*b*) the blade angle at the outlet edge of the impeller.

Assume that the manometric efficiency is 75 per cent; the loss of head in the pump due to fluid resistance is $0·033v^2$ m, where v m/s is the absolute velocity with which the water is discharged from the impeller; the area of the impeller outlet surface is $1·2D^2$ m² where D is the impeller diameter in m; and water enters the impeller without whirl.
Answer 0·253 m, 30°

7 A centrifugal pump is required to discharge 0·56 m³/s of water and develop a head of 12 m when the impeller rotates at

750 rev/min. The manometric efficiency is to be 80 per cent, the loss of head in the pump due to friction being assumed to be $0 \cdot 0276 V^2$ m of water, where V is the velocity with which water leaves the impeller. Water enters the impeller without shock or whirl and the velocity of flow is constant at $2 \cdot 7$ m/s. Obtain (a) the impeller diameter and outlet area, and (b) the blade angle at the outlet edge of the impeller. Explain briefly why the direction of the actual velocity at discharge from the impeller differs usually from the direction given by the outlet velocity diagram.

Answer $0 \cdot 364$ m, $0 \cdot 207$ m², 34°

8 A centrifugal pump with an impeller of 190 mm diam gives at maximum efficiency a discharge of $3 \cdot 9$ m³/min of fresh water at 1800 rev/min against a head of $4 \cdot 2$ m. What should be the speed of rotation of a similar impeller of 380 mm diam to give $54 \cdot 5$ m³/min of sea water of density 1025 kg/m³ and what pressure would it then generate?

Answer 3150 rev/min, 515 kN/m³

9 A scale model, one-fifth full size, is to be tested in order to predict the performance of a large centrifugal pump working against a head H. Show that, provided the viscosity of the fluid has no appreciable effect on the performance of the pump, the test may be carried out under any convenient head.

What head would be required for the test if viscosity is taken into account, (a) when the pump and model both use water, (b) when the kinematic viscosity of the fluid dealt with by the pump is 5 times that used by the model, and what would be the ratio of the corresponding speeds of rotation in each case? Establish the formulae required.

Answer (a) $H_m = 25H$, $N_m = 25N$; (b) $H_m = H$, $N_m = 5N$

10 A centrifugal fan has to deliver $4 \cdot 25$ m³/s when running at 750 rev/min. The diameter of the impeller at inlet is 525 mm and at outlet is 750 mm. It may be assumed that the air enters radially at a speed of 15 m/s. The vanes are set backwards at outlet at 70° to the tangent and the width at outlet is 100 mm. The volute casing gives a 30 per cent recovery of the outlet velocity head. The losses in the impeller may be taken as equivalent to 25 per cent of the outlet velocity head. The specific volume of air is $0 \cdot 8$ m³/kg and the blade thickness effects may be neglected. Determine the manometric efficiency, the power required and the pressure at discharge.

Answer $57 \cdot 9$ per cent, $2 \cdot 08$ kW, $39 \cdot 2$ m of air

11 A centrifugal pump impeller is of 250 mm external diameter and the water passage is 32 mm wide at exit. The circumference is reduced by 12 per cent on account of vane thickness. The impeller vanes are inclined at 140 deg to the forward tangent at exit. Manometric efficiency = 83 per cent, rev/min = 1000, $Q = 2 \cdot 86$ m³/min. Calculate the conversion efficiency of the diffuser ring. Assume no losses in the impeller.

Answer 56 per cent

12 The impeller of a centrifugal pump has an external diameter of 250 mm and an effective outlet area of 170 cm². The blades are bent back so that the angle at the outlet edge is 148 deg to the tangent drawn to the direction of impeller rotation. The diameters of the suction and delivery openings are 150 mm and 125 mm respectively.

When running at 1450 rev/min and delivering 28 dm³/s of water, the pressure heads at the suction and delivery openings were found to be respectively 4·5 m *below* and 13·5 m *above* atmospheric pressure, the points at which these pressure heads were measured being at the same level. The motor driving the pump supplied 8 kW. Water enters the impeller without shock or whirl.

Assuming that the true outlet whirl component = 0·7 of the ideal one, obtain: (*a*) the overall efficiency; (*b*) the manometric efficiency based on the true whirl component.
Answer (*a*) 61·4 per cent; (*b*) 83·4 per cent

13 A centrifugal pump delivers 11·8 cubic metres of water per minute at 1200 rev/min with a manometric efficiency of 75 per cent. The impeller is 300 mm diam with a width at exit of 75 mm. The blades occupy 12 per cent of the periphery and are swept backwards making an angle of 40 deg with the tangent at the outer periphery. Calculate the fraction *k* of the kinetic energy of discharge from the impeller which is subsequently recovered in the casing assuming no loss of head in the impeller. What is the manometric efficiency if *k* = 0?

Make simple sketches to show two conventional methods used to regain the kinetic energy of discharge from the impeller.
Answer 0·402, 58 per cent

14 Derive the expression for the specific speed of a centrifugal pump in terms of *N* the speed of impeller rotation, *Q* the quantity discharged and *H* the head developed.

A multi-stage centrifugal pump having 6 stages with 225 mm diam impellers develops a head of 120 m when running at 1500 rev/min and discharging 5·45 cubic metres of water per minute.

Four geometrically similar stages having 300 mm diam impellers are used to build a multi-stage pump which is to run at 1000 rev/min. Assuming that each stage in both pumps operates under dynamically similar conditions, obtain (*a*) the quantity in m³/min that will be discharged by this pump and (*b*) the head that it will develop.
Answer (*a*) 8·61 m³/min, (*b*) 63·2 m

15 A centrifugal pump produced the following performance data when running at 1500 rev/min on a test run.

Flow m³/s	0·075	0·150	0·200	0·250	0·300
Total head m	70	68	64	58	49
Input power kW	97	127	147	163	170

The pump is required to deliver water from a sump to a reservoir whose level is 60 m above that of the sump. Suction and delivery pipes of 300 mm diam will have a combined length of 120 m ($f = 0 \cdot 006$), 12 m of which is on the suction side, and the pump inlet is 3 m above the water level in the supply sump. What will be the efficiency and the discharge of the pump at the test speed? What would be the most economical speed to operate the pump and what suction head would occur at the pump inlet under these optimum speed conditions?

Answer   85·3 per cent, 0·2 m³/s; 1620 rev/min, 4·3 m

16   Define the term "specific speed" of a centrifugal pump and deduce an expression for it in terms of the head H, the discharge Q and the speed N.

A multi-stage centrifugal pump is required to lift 1·8 m³/min of water from a mine, the total head including friction being 750 m. If the speed of the pump is 2900 rev/min, find the least number of stages if the specific speed per stage is not to be less than 150 in SI units.

Answer   10 stages

15

Reciprocating pumps

The reciprocating mechanism consists of a piston or displacer moving in a cylinder which liquid enters or leaves through suitable valves. The piston is given a reciprocating motion by means of a connecting rod and crank.

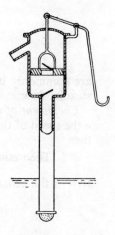

Figure 15.1

Suction pumps (Fig. 15.1) are used solely to raise water to the pump cylinder level. On the suction stroke the movement of the piston forms a partial vacuum in the cylinder and the atmospheric pressure forces the liquid in the sump into the cylinder. Theoretically the lift cannot exceed the head of liquid equivalent to atmospheric pressure, which in the case of water is 10·4 m, but if the pressure falls below the vapour pressure the liquid will boil in the cylinder and the pump cease to function. The available lift in the case of water is thus limited to about 0·8 m at ordinary temperatures.

Force pumps (Fig. 15.2) are similar to suction pumps but on the delivery stroke the liquid is forced into a delivery pipe and can be raised to any desired height above the pump centre-line. The same limitations on the lift from sump to pump cylinder apply as for the suction pump.

Single-acting pumps make one delivery per revolution of the crank for each cylinder (Fig. 15.2).

Double-acting pumps (Fig. 15.3) make two deliveries per revolution of the crank for each cylinder.

A multi-cylinder pump has two or more cylinders. Three cylinders

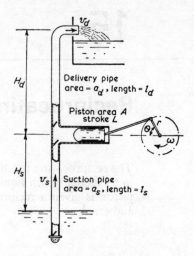

Figure   15.2

are commonly used with the three cranks set 120 deg apart giving a steadier flow.

Coefficient of discharge. If A is the area of the piston, L the stroke, n the speed of the crank in rev/s and w the specific weight of the liquid, then

$$\text{Theoretical weight discharged per sec} = wALn$$

If W is the actual weight discharged per sec

$$\text{Coefficient of discharge} = \frac{W}{wALn}$$

and is usually less than unity owing to leakage.

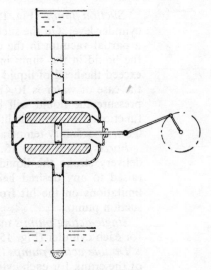

Figure   15.3

Alternatively the volumetric performance of the pump is given by

$$\text{Percentage slip} = \frac{\text{volume swept} - \text{volume discharged}}{\text{volume swept}} \times 100$$

When the delivery head is low and there is a long suction pipe, the inertia of the liquid may cause the delivery valve to open early so that the liquid flows straight through the cylinder giving a discharge greater than the swept volume, a coefficient of discharge greater than unity and a negative slip.

Separation occurs when the pressure at the face of the piston falls below vapour pressure and the liquid is separated from the piston by vapour.

15.1 Theoretical indicator diagram, effect of acceleration and friction

Give a sketch of the theoretical pressure/volume diagram for the cylinder of a reciprocating pump, which is not fitted with air vessels. Show clearly the effect of acceleration and friction in both suction and delivery pipes. State the conditions under which "separation" is likely to occur.

The following particulars relate to a reciprocating pump not fitted with air vessels: stroke, $0 \cdot 3$ m; plunger diameter, 125 mm; diameter of suction pipe, 75 mm; length 6 m; suction head, 3 m. Atmospheric pressure is equivalent to $10 \cdot 2$ m of water and separation may be assumed to occur when the absolute pressure head in the cylinder falls below $2 \cdot 4$ m of water. Calculate the maximum speed at which the pump may be run if separation is to be avoided.

Solution If H_s = section head (Fig. 15.2) and H_d = delivery head and the hydraulic effects of the pipeline are ignored, the theoretical pressure/volume diagram or indicator diagram would be as shown in Fig. 15.4(*a*), the pressure being constant at H_s during the suction stroke and H_d during delivery stroke.

Area *abcd* = work done on suction stroke

Area *defa* = work done on delivery stroke

The hydraulic effects of the suction and delivery pipes will modify this diagram as a result of (*a*) the effects of acceleration and retardation of flow in the pipes, (*b*) the friction losses in the pipes.

For a single-cylinder single-acting pump the effects can be calculated as follows.

(*a*) The flow in the suction pipes is fluctuating from zero at the beginning of the suction stroke to a maximum at mid-stroke and back to zero at the end of the stroke. Additional suction head H_{as} is required to accelerate the flow at the beginning of the stroke and an equal and opposite head is required to bring the flow to rest at the end of the stroke. Similarly during the delivery stroke there will be an acceleration head H_{ad} for the delivery pipe.

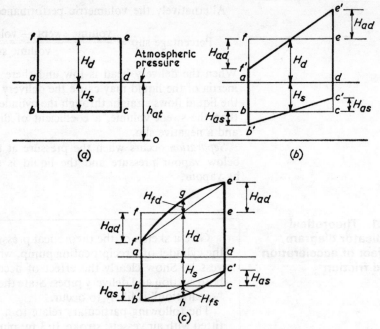

(a)

(b)

(c)

Figure 15.4

Assuming simple harmonic motion of the piston, from Fig. 15.5 if ω is the angular velocity of the crank and θ is its displacement from dead centre, $\theta = \omega t$ and

$$\text{Displacement of piston} = x = r - r\cos\theta$$

$$= r(1 - \cos\omega t)$$

$$\text{Velocity of piston} = V = \frac{dx}{dt} = \omega r\sin\omega t$$

For either pipe (suction or delivery) of cross-sectional area a,

$$\text{Velocity of flow in pipe} = v = \frac{A}{a}V = \frac{A}{a}\omega r\sin\omega t = \frac{A}{a}\omega r\sin\theta$$

$$\text{Acceleration of liquid in pipe} = \frac{dv}{dt} = \frac{A}{a}\omega^2 r\cos\omega t = \frac{A}{a}\omega^2 r\cos\theta$$

$$\text{Mass of liquid in pipe} = \rho a l$$

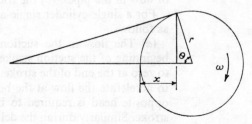

Figure 15.5

If H_a is the acceleration head required to produce this acceleration

$$\text{Force due to } H_a = \rho g H_a a$$

and since force = mass × acceleration

$$\rho g H_a a = \rho a l \times \frac{A}{a} \omega^2 r \cos \theta$$

$$H_a = \frac{l}{g} \cdot \frac{A}{a} \cdot \omega^2 r \cos \theta$$

The displacement of the piston also depends on cos θ so that if H_a is plotted on the indicator diagram it will give a straight line.

For the suction stroke:

At beginning of stroke $\quad \theta = 0$ and $H_{as} = \dfrac{l_s}{g} \dfrac{A}{a_s} \omega^2 r$

At mid-stroke $\quad \theta = 90°$ and $H_{as} = 0$

At end of stroke $\quad \theta = 180°$ and $H_{as} = - \dfrac{l_s}{g} \dfrac{A}{a_s} \omega^2 r$

Plotting these on the original indicator diagram, the modified diagram is shown in Fig. 15.4(b). The acceleration effect requires an increase of suction from ab to ab' at the beginning of the stroke and a reduction from dc to dc' at the end of stroke. The increased suction ab' increases the tendency to separation and since H_a depends on ω there will be a maximum speed above which separation will occur.

Similarly on the delivery stroke the pressure is increased by

$$H_{ad} = \frac{l_d}{g} \frac{A}{a_d} \omega^2 r$$

at the beginning of the stroke and reduced at the end of the stroke. Note that the area of the indicator diagram and therefore the work done is unchanged.

(b) The loss of head in the appropriate pipe (suction or delivery) at any point in the stroke depends on the velocity v and is calculated from the Darcy formula.

$$H_f = \frac{4fl}{d} \frac{v^2}{2g} = \frac{4fl}{2gd} \left(\frac{A}{a} \omega r \sin \theta \right)^2$$

Plotted on a stroke basis this gives a parabolic variation.
For the suction stroke:

At the beginning and end $\quad \sin \theta = 0 \quad$ and $\quad H_{fs} = 0$

At mid-stroke $\quad \theta = 90° \quad$ and $\quad H_{fs} = \dfrac{4fl_s}{2gd_s} \left(\dfrac{A}{a_s} \omega r \right)^2$

Similarly on the delivery stroke the maximum loss occurs at mid-stroke where $H_{fd} = \dfrac{4fl_d}{2gd_d} \left(\dfrac{A}{a_d} \omega r \right)^2$.

Work is done against friction and since the mean ordinate of a parabola is two-thirds of the maximum, then if W is weight per unit time discharged:

Work done against friction

$$\text{on suction stroke} \qquad \frac{2}{3} \frac{4fl_s}{2gd_s} \left(\frac{A}{a_s} \omega r \right)^2 \times W$$

$$\text{on delivery stroke} \qquad \frac{2}{3} \frac{4fl_d}{2gd_d} \left(\frac{A}{a_d} \omega r \right)^2 \times W$$

Figure 15.4(c) shows the complete theoretical indicator diagram including the effects of acceleration and friction.

$$\text{Total work done} = W(H_s + H_d + \tfrac{2}{3}H_{fs} + \tfrac{2}{3}H_{fd})$$

Separation will occur at the beginning of suction stroke if pressure head in cylinder is less than 2·4 m absolute.

$$\text{Pressure head in cylinder} = H_{at} - H_s - H_{as} - H_{fs}$$

$$H = H_{at} - H_s - \frac{l_s}{g} \frac{A}{a_s} \omega^2 r = 0$$

Putting $H_{as} = 10 \cdot 2\,\text{m}$, $H_s = 3\,\text{m}$, $H = 2 \cdot 4\,\text{m}$, $l_s = 6\,\text{m}$, $\dfrac{A}{a_s} = \left(\dfrac{125}{75} \right)^2$

$$= \frac{25}{9}, r = \frac{1}{2} \text{ stroke} = 0 \cdot 15\,\text{m}$$

$$8 = 34 - 10 - \frac{6}{g} \times \frac{25}{9} \times \omega^2 \times 0 \cdot 15$$

$$\omega^2 = 18 \cdot 8, \quad \omega = 4 \cdot 33\,\text{rad/s}$$

$$\text{Maximum speed} = \frac{\omega \times 60}{2\pi} = 41 \cdot 4\,\text{rev/min}$$

15.2 Use of air vessels

Explain the reason for fitting large air vessels on the suction and delivery pipes of a reciprocating pump close to the cylinder.

A reciprocating pump, working with simple harmonic motion, may be fitted with large air vessels, near the pump cylinder, in both the suction and delivery pipes. Working from first principles show that, in the case of a single-acting pump, the ratio of the work done against pipe friction, if air vessels are fitted, compared with that done if there are no air vessels, is $3/(2\pi^2)$. It is assumed that the friction coefficient does not vary with velocity.

Solution. The purpose of fitting large air vessels close to the cylinder is to smooth out the flow in the suction and delivery pipes so that flow is continuous in both pipes during both suction and delivery strokes. On the delivery stroke when the suction valve is closed the flow in the suction pipe passes into the air vessel (vacuum chamber) on the suction pipe, Fig. 15.6. Similarly on the suction stroke the flow in the delivery pipe is maintained by liquid discharged by the air under pressure from the air vessel on the delivery side which receives part of the discharge from the cylinder on the delivery stroke.

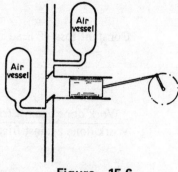

Figure 15.6

It is usual to assume that, if large air vessels are fitted, the pipe velocities are constant.

$$\text{Velocity in suction pipe} = \frac{Q}{a_s}$$

$$\text{Velocity in delivery pipe} = \frac{Q}{a_d}$$

where Q = discharge, a_s = area of suction pipe, a_d = area of discharge pipe.

Since the velocity is constant there is no acceleration head in the pipes except for a short length between the air vessel and cylinder. This reduces the risk of separation and allows the pump to operate at higher speeds and increased discharge.

Also since discharge is continuous the velocity in the pipes is reduced and therefore the head lost in friction is reduced. For a single-acting pump:

When no air vessel is fitted

From example **15.1**, assuming simple harmonic motion

$$\text{Velocity of flow in pipe} = v = \frac{A}{a}\,\omega r \sin\theta$$

$$\text{Maximum loss of head in friction} = \frac{4fl}{2gd}\left(\frac{A}{a}\,\omega r\right)^2$$

and since the frictional part of the indicator diagram, Fig. 15.4(c), is a parabola

$$\text{Work done against friction per unit wt} = \frac{2}{3}\times\frac{4fl}{2gd}\left(\frac{A}{a}.\omega r^2\right)$$

When air vessels are fitted

$$\text{Discharge} = \text{piston area} \times \text{stroke} \times \text{rev/sec}$$

$$Q = 2rAn = 2rA\frac{\omega}{2\pi}$$

$$\text{Velocity of flow in pipe} = \frac{Q}{a} = \text{constant}$$

$$= \frac{A}{a}\frac{\omega r}{\pi}$$

$$\text{Constant loss of head in friction} = \frac{4fl}{2gd}\left(\frac{A}{a}\frac{\omega r}{\pi}\right)^2$$

$$= \text{Work done against friction per unit wt}$$

$$\frac{\text{Work done against friction with air vessels}}{\text{Work done against friction without air vessels}} = \frac{\left(\frac{1}{\pi}\right)^2}{\frac{2}{3}} = \frac{3}{2\pi^2}$$

15.3 Double-acting pump

A double-acting reciprocating pump of 175 mm diam by 350 mm stroke draws from a source 3 m below, and delivers to a height 46 m above, its own level. Both suction and delivery pipes are 100 mm diam and their respective lengths are 6 m and 75 m. The pump piston has simple harmonic motion and makes 40 double strokes per min. Large air vessels are fitted on both sides of the pump. The air vessel on the suction side is $1 \cdot 5$ m away from the cylinder, while that on the delivery side is $4 \cdot 5$ m away. The friction coefficient for the pipes is $0 \cdot 008$. Determine the pressure difference between the two sides of the piston at the beginning of the stroke.

Solution. At the beginning of the stroke the pressure on one side of the piston is the maximum for delivery and on the other side the maximum suction, each being composed of the lift + acceleration head for the length of pipe between air vessel and cylinder + friction head for uniform flow in remainder of pipe + velocity head.

If the pump makes 40 double strokes per minute then, since it is double-acting

Discharge Q

$$= \text{area} \times \text{stroke} \times 2 \times \text{rev/s}$$

$$= \tfrac{1}{4}\pi \times 0 \cdot 175^2 \times 0 \cdot 35 \times 2 \times 40/60 = 11 \cdot 22 \times 10^{-3}\,\text{m}^3/\text{s}$$

Area of suction pipe $= a_s = $ area of delivery pipe $= a_d$

$$= \tfrac{1}{4}\pi \times 0 \cdot 1^2 = 7 \cdot 85 \times 10^{-3}\,\text{m}^2$$

Pipe velocity $= v_s = v_d = \dfrac{Q}{\text{area}}$

$$= \frac{11 \cdot 22 \times 10^{-3}}{7 \cdot 85 \times 10^{-3}} = 1 \cdot 43 \, \text{m/s}$$

For the suction side

$$\text{Suction head} = H_s = 3 \, \text{m}$$

From example **15.1**, considering length of pipe l_g between air vessel and cylinder:

$$\text{Acceleration head} = H_{as} = \frac{l_s}{g} \frac{A}{a_s} \omega^2 r$$

$$l_s = 1 \cdot 5 \, \text{m}, \quad \frac{A}{a_s} = \left(\frac{175}{100} \right)^2 = \frac{49}{16}, \quad \omega = 2\pi \times \frac{40}{60} = 4 \cdot 189 \, \text{rad/s}$$
$$r = 0 \cdot 175 \, \text{m},$$

$$H_{as} = \frac{1 \cdot 5}{g} \times \frac{49}{16} \times (4 \cdot 189)^2 \times 0 \cdot 175 = 1 \cdot 44 \, \text{m}$$

$$\text{Velocity head} = \frac{v_s^2}{2g} = \frac{1 \cdot 43^2}{19 \cdot 62} = 0 \cdot 104 \, \text{m}$$

Friction head for the length of pipe L_s up to air vessel

$$= H_{fs} = \frac{4fL_s}{d_s} \frac{v_s^2}{2g}$$

$$= \frac{4 \times 0 \cdot 008 \times (6 - 1 \cdot 5)}{0 \cdot 1} \times 0 \cdot 104 = 0 \cdot 15 \, \text{m}$$

$$\text{Total suction head} = H_s + H_{as} + H_{fs} + \frac{v_s^2}{2g}$$

$$= 3 + 1 \cdot 44 + 0 \cdot 15 + 0 \cdot 1 = 4 \cdot 69 \, \text{m}$$

For the delivery side

Delivery head $= H_d = 46 \, \text{m}$

$$\text{Acceleration head} = H_{ad} = \frac{l_d}{g} \frac{A}{a_d} \cdot \omega^2 r$$

$$= \frac{4 \cdot 5}{g} \times \frac{49}{16} \times (4 \cdot 189)^2 \times 0 \cdot 175 = 4 \cdot 32 \, \text{m}$$

$$\text{Velocity head} = \frac{v_d^2}{2g} = \frac{1 \cdot 43^2}{19 \cdot 62} = 0 \cdot 104 \, \text{m}$$

Friction head for pipe beyond air vessel

$$= H_{fd} = \frac{4fL_d}{d_d} \frac{v_d^2}{2g}$$

$$= \frac{4 \times 0 \cdot 008 \times (75 - 4 \cdot 5)}{0 \cdot 1} \times 0 \cdot 104 = 2 \cdot 38 \, \text{m}$$

$$\text{Total delivery head} = H_d + H_{ad} + H_{fd} + \frac{v_d{}^2}{2g}$$

$$= 46 + 4 \cdot 32 + 2 \cdot 38 + 0 \cdot 10 = 52 \cdot 80\,\text{m}$$

Pressure head difference between suction and delivery side

$$= 4 \cdot 69 + 52 \cdot 80 = \mathbf{57 \cdot 49\,m}$$

Problems

1 A double-acting reciprocating pump has a piston diam of 200 mm and a stroke of 0·6 m and runs at 20 rev/min. It discharges through a 150 mm main 75 mm long ($f = 0·0075$) with a vertical lift of 45 m. Assuming the piston to have s.h.m. and that no air vessel is used, sketch the part of the indicator card corresponding to discharge giving the heads in the cylinder at the ends and middle of the stroke. Neglect friction at the discharge valve.
Answer 62·9 m, 45·96 m, 27·1 m

2 The plunger diam of a single-acting reciprocating pump is 115 mm and the stroke is 230 mm. The suction pipe is 90 mm diam and 4·2 m long. If separation takes place at an absolute pressure of 1·2 mm find the maximum speed at which the pump will run without separation taking place if the barometer stands at 10·3 m of water and the water level in the sump is 3 m below the pump cylinder axis. What power is expended in overcoming friction at this speed, taking $f = 0·01$?
Answer 83·5 rev/min, 5·5 W

3 A single-acting reciprocating pump has a plunger diam of 250 mm and a stroke of 450 mm. The delivery pipe is 110 mm diam and 48 m long. If the plunger moves with s.h.m. find the power saved in overcoming friction in the delivery pipe by the provision of a large air vessel on this pipe close to the cylinder when the pump is driven at 20 rev/min, taking $f = 0·01$.
Answer 215 W

4 A double-acting reciprocating pump is used to raise water to a height of 42 m through a delivery pipe of diam 75 mm and length 81 m. The pump speed is 180 rev/min, the stroke is 250 mm and the piston diam is 115 mm. A large air vessel is fitted in the delivery pipe at 6 m from the cylinder, measured along the pipe.

Determine the absolute pressure in the cylinder at the end of each delivery stroke, given that the friction coefficient for the pipe is 0·007. It should be assumed that the piston moves with simple harmonic motion, that the effect of the piston rod is negligible, and that the atmospheric pressure is 10·2 m of water.
Answer 6·94 m of water

5 A double-acting single-cylinder reciprocating pump of 190 mm bore and 380 mm stroke runs at 36 double strokes/min, suction head 3·6 m, and discharge head 30 m. The length of the suction pipe is 9 m and of the discharge pipe 60 m, and the diameter of each pipe 100 mm. Large air vessels are provided 3 m away from the pump on the suction side, and 6 m away on the discharge side, both measured along the pipelines; $f = 0·008$.

Neglecting entrance and exit losses for the pipes, estimate for the beginning of the stroke: (*a*) the head in the two ends of the cylinder, (*b*) the load on the piston rod, neglecting the size of the piston rod and assuming simple harmonic motion.

Answer   34·04 m, −4·75 m, 10·8 kN

6 A single-acting reciprocating pump has a piston of 200 m diam and 600 mm stroke. It runs at 20 rev/min, motion being simple harmonic. Delivery is through a 100 mm diam pipe of length 45 m for which $f = 0·008$. Find the power which would be saved by fitting an air vessel to the delivery side assuming that there would then be no acceleration in the pipe.

Answer   66 W

7 Sketch theoretical indicator diagrams for a single-cylinder single-acting reciprocating pump not fitted with air vessels. Use your diagram to explain clearly the effect of acceleration and friction on both suction and delivery strokes.

Assuming simple harmonic motion of the piston, develop an expression for the acceleration head in the cylinder at the beginning of the suction stroke in such a pump.

The following data relate to a pump of the type described above: length of suction pipe, 9 m; diameter of suction pipe, 75 mm; suction lift, 3 m; plunger diameter, 125 mm; stroke, 300 mm; speed, 30 rev/min. Calculate the theoretical absolute pressure head in metres of water at the beginning and end of the suction stroke. Barometric pressure corresponds to 10·2 m of water.

Answer   3·43 m, 10·97 m

8 A reciprocating pump has three single-acting cylinders the pistons of which are operated by cranks at 120 deg apart. The pistons are 75 mm diam and have a stroke of 150 mm.

The cylinders discharge water into a single pipe 50 mm diam and 60 m long. The pump speed is 60 rev/min and there is no air vessel on the delivery side.

Give diagrams on a crank angle base showing how the water velocity and acceleration in the discharge pipe differ from those obtained with a single-acting single-cylinder pump of the above dimensions and speed.

If the pipe discharges to air 30 m above the level of the cylinders calculate the range of pressure in the pipe just beyond the pump. Take f for the pipe as 0·01.

Answer   52·46 to 11·66 m of water

9 A reciprocating pump has two cylinders and the two cranks are at 180 deg. The pistons are single-acting 40 mm diam with a stroke of 150 mm. The pump runs at 80 rev/min and delivers directly into an accumulator which has a ram 100 mm diam loaded with dead weights having a mass of 8000 kg. The ram friction amounts to 900 N.

If the total inflow to the accumulator per minute is just balanced by the outflow, which is quite uniform, calculate the pressure range in the pump cylinder and give a sketch of the theoretical indicator diagram, assuming a very short suction pipe. Assume s.h.m. for the pistons.

Answer 1716 kN/m²

10 Explain the object of fitting an air vessel on (*a*) the suction side and (*b*) the delivery side of a reciprocating pump.

A single-acting reciprocating pump with plunger diam of 100 mm and stroke of 150 mm has a speed of 75 rev/min. The centre of the pump cylinder is 1·5 m above the level of the water in the sump. The 75 mm diam suction pipe is 7·2 m long. The level of the delivery tank is 30 m above the centre of the pump cylinder, and the 63 mm diam delivery pipe is 75 m long. The friction coefficient for the suction and delivery pipes is 0·01. There is no air vessel on the suction side, but one, which may be assumed to be perfectly efficient, is installed on the delivery side.

Assuming that the plunger moves horizontally with simple harmonic motion, determine (*a*) the pressure on the plunger at the beginning, middle and end of the suction stroke; (*b*) the water power of the pump. Also obtain the pressure on the plunger at the beginning of the delivery stroke if no air vessel had been fitted on the delivery side.

Answer (*a*) −7·56 m, −1·71 m, +4·56 m of water,
(*b*) 465 W, 119 m of water

11 A reciprocating pump has a cylinder of 75 mm bore × 150 mm stroke and draws water from a sump whose level is 1·5 m below the axis of the pump. If the suction pipe is 2·4 m long and 50 mm in diameter, find the speed of the pump in rev/min at which separation occurs if this takes place at a vacuum head of 7·9 m of water. Assume simple harmonic motion of the piston. If *f* = 0·01 for the pipe, what is the friction head at mid-stroke when running at this speed?

Answer 119 rev/min, 0·435 m

12 A double-acting single-cylinder reciprocating pump has a cylinder diameter of 150 mm and 450 mm stroke. The suction and delivery pipes have diameters and lengths of 100 mm, 6 m and 75 mm, 60 m respectively. The sump is 4·5 m below and the reservoir 45 m above the centre-line of the pump. If the pump runs at 60 rev/min, determine the power of the driving motor if its efficiency is 0·85. Take *f* = 0·005 and assume s.h.m. for the plunger.

Answer 12·6 kW

16

Hydraulically-operated machinery

Shaping presses, cranes, lifts and other machinery are operated by hydraulic mechanisms of the piston and ram type, which are basically similar to the hydraulic jack (see Volume 1), but the solution of any particular problem usually involves the action of several units including of course the hydraulic behaviour of the connecting pipes.

Figure 16.1 shows diagrammatically a typical circuit. Low-pressure

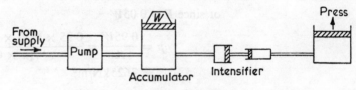

Figure 16.1

liquid (oil or water) enters the pump and is delivered at high pressure into an accumulator where it is stored and drawn off as required. If the accumulator pressure is insufficient a higher pressure can be obtained by means of an intensifier.

The *accumulator* consists of a storage cylinder in which the fluid is kept under pressure either by means of a dead load carried on a moving piston or by means of air pressure in a closed pressure vessel.

The *intensifier* can be regarded as a low-pressure cylinder with a large piston which is coupled mechanically to a small piston working in the high-pressure cylinder. Note that there is no hydraulic connexion between the low- and high-pressure sides.

For cranes or lifts the comparatively small movement of the piston in the working cylinder is magnified by a system of ropes and pulleys which may form part of the piston and cylinder arrangement which is then called a *jigger*.

16.1 Accumulator

A hydraulic accumulator has a 450 mm diam ram and 7 m lift and carries a load which has a mass of 130 metric tons. If friction accounts for 5 per cent of the total pressure on the ram, determine the power given into the delivery main if the accumulator empties steadily in 3 min and if at the same time the pumps are supplying 7·58 dm³/s to the accumulator.

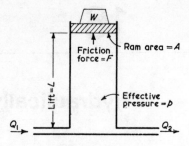

Figure 16.2

Solution Fig. 16.2 shows the arrangement. Let W = load on accumulator ram, A = area of ram, F = frictional resistance, p = effective pressure in accumulator.

$$p = \frac{\text{effective force}}{\text{ram area}} = \frac{W - F}{A}$$

or since $F = 0.05W$

$$p = \frac{0.95W}{A} = \frac{0.95 \times (130 \times 1000 \times 9.81)}{\frac{1}{4}\pi \times 0.45^2} \, \text{N/m}^2$$
$$= 7625 \, \text{kN/m}^2$$

If Q_2 is the combined discharge from the accumulator and pumps and Q_1 = input from pumps

$$\text{Discharge from accumulator } Q = \frac{\text{volume}}{\text{time of emptying}}$$
$$= \frac{\text{area} \times \text{lift}}{\text{time of emptying}}$$

$$Q = \tfrac{1}{4}\pi \times 0.45^2 \times \frac{7}{3 \times 60} = 6.19 \times 10^{-3} \, \text{m}^3/\text{s}$$

$$Q_1 = 7.58 \, \text{dm}^3/\text{s} = 7.58 \times 10^{-3} \, \text{m}^3/\text{s}$$

$$Q = Q + Q_1 = 6.19 \times 10^{-3} + 7.58 \times 10^{-3}$$
$$= 13.77 \times 10^{-3} \, \text{m}^3/\text{s}$$

$$\text{Power available} = pQ_2 = (7625 \times 10^3) \times (13.77 \times 10^{-3}) \text{W}$$
$$= \mathbf{105 \, kW}$$

16.2 Intensifier and ram

An intensifier of the piston and ram type has a low-pressure water supply with a head of 18 m. The diameters of the intensifier piston and ram are 900 mm and 75 mm respectively and the diam of the ram operated is 200 mm. The low-pressure supply is 50 mm diam, 105 m long, $f = 0.006$. Take the friction loss at each packing as 5 per cent of the pressure on the ram or piston concerned, neglect other losses and find the speed of the press ram in millimetres/minute when it is exerting a force of 600 000 N.

Solution In Fig. 16.3 v_1, L, d are the velocity, length and diameter of the supply pipe; p_2, A_2, v_2 are the pressure, piston area and piston velocity on the low-pressure side of the intensifier; p_3, A_3, v_3 are the corresponding quantities on the high-pressure side; p_4, A_4 and V are the values for the press ram.

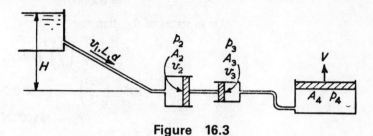

Figure 16.3

For the supply pipe, consider friction only,

$$\frac{p_2}{w} = H - \frac{4fL}{d}\frac{v_1^2}{2g} \tag{1}$$

For continuity of flow on the low pressure side, if A_1 is the pipe area

$$A_1 v_1 = A_2 v_2$$

For continuity of flow on the high-pressure side

$$A_3 v_3 = A_4 V$$

For the intensifier

$$v_2 = v_3$$

Thus

$$v_1 = \frac{A_2}{A_1} v_2 = \frac{A_2}{A_1} v_3 = \frac{A_2}{A_1} \times \frac{A_4}{A_3} V$$

or in terms of the corresponding diameters

$$v_1 = \left(\frac{d_2}{d_1}\right)^2 \times \left(\frac{d_4}{d_3}\right)^2 \times V$$

Putting $d_1 = 50\,$mm, $d_2 = 900\,$mm, $d_3 = 75\,$mm, $d_4 = 200\,$mm

$$v_1 = \left(\frac{900}{50}\right)^2 \times \left(\frac{200}{75}\right)^2 V = 2304V \tag{2}$$

Allowing for 5 per cent friction loss at each packing,

Available force on low-pressure piston of intensifier

$$= 0.95 A_2 p_2$$

Available force exerted by high-pressure piston of intensifier

$$= (0.95)^2 A_2 p_2$$

$$p_3 = \frac{\text{force}}{\text{area}} = (0.95)^2 \frac{A_2}{A_3} p_2$$

Also

$$p_3 = p_4$$

Available force exerted by press ram

$$= 0.95A_4p_4$$

$$= 0.95A_4p_3 = (0.95)^3 \frac{A_2 \times A_4}{A_3} p_2$$

$$= 600000\,\text{N}$$

Or in terms of the diameters

$$(0.95)^3 \left(\frac{d_2}{d_3}\right)^2 \tfrac{1}{4}\pi d_4{}^2 p_2 = 600000$$

$$(0.95)^3 \left(\frac{900}{75}\right)^2 \times \tfrac{1}{4}\pi \times 0.2^2 \times p_2 = 600000$$

$$p_2 = 155000\,\text{N/m}^2 \qquad (3)$$

Substituting from equations (2) and (3) in (1)

$$\frac{155000}{9.81 \times 1000} = 18 - \frac{4 \times 0.006 \times 105 \times (2304V)^2}{0.05 \times 2g}$$

$$15.8 = 18 - 13.7 \times 10^6 V^2$$

$$V = 0.401 \times 10^{-3}\,\text{m/s} = \textbf{24\,mm/min}$$

16.3 Hydraulic crane

A hydraulic crane is operated by a ram fitted with a jigger (a device for multiplying the movement of the ram). The crane is required to lift a load of 6 metric tons at 1·2 m/s through 78 m once every 5 min. The ram diam is 750 mm, the water pressure is 3950 kN/m² and the mechanism may be assumed to be 83 per cent efficient. A pump supplies water continuously to the ram, via an accumulator which can store water at the required ram pressure. Determine (a) the required minimum output power of the pump, (b) the minimum volumetric capacity of the accumulator, and (c) the stroke of the jigger ram.

Solution. (a) Work done per lift = load × lift

$$= 6 \times 10^3 \times 9.81 \times 78 = 4.6 \times 10^6\,\text{N-m}$$

Time between lifts = 5 min = 300 s

∴ Power required from pump if run continuously is

$$\frac{\text{work done per lift}}{\text{efficiency} \times \text{time between lifts}} = \frac{4.6 \times 10^6}{0.83 \times 300}\,\text{W}$$

$$= \textbf{18.4\,kW}$$

(b) If Q = rate of discharge of pump at pressure p

Water power = $pQ = 18.4\,\text{kW} = 18.4 \times 10^3\,\text{W}$

Putting $p = 3950 \times 10^3\,\text{N/m}^2$

$$Q = \frac{18 \cdot 4 \times 10^3}{3950 \times 10^3} = 4 \cdot 67 \times 10^{-3}\,\mathrm{m^3/s}$$

Velocity of lift $= 1 \cdot 2\,\mathrm{m/s}$, Lift $= 78\,\mathrm{m}$

Time of lifting, during which accumulator is not filling $= \dfrac{78}{1 \cdot 2}$

$$= 65\,\mathrm{s}$$

Interval between lifts $= 5\,\mathrm{min} = 300\,\mathrm{s}$

Time during which accumulator is filling $= 300 - 65 = 235\,\mathrm{s}$

Minimum capacity $= 235Q = 235 \times 4 \cdot 67 \times 10^{-3}\,\mathrm{m^3} = \mathbf{1 \cdot 1\,m^3}$

(c) Total flow from pumps in $5\,\mathrm{min} = 300 \times 4 \cdot 65 \times 10^{-3} = 1 \cdot 39\,\mathrm{m^3}$
The whole of this volume is discharged to the jigger ram.

$$\text{Stroke} = \frac{\text{volume}}{\text{area}} = \frac{1 \cdot 39}{\frac{1}{4}\pi \times 0 \cdot 75^2} = \mathbf{3 \cdot 15\,m}$$

Problems

1 A hydraulic crane of 150 mm ram diam is supplied with water from an accumulator, loaded with 100 metric tons and having a ram diam of 550 mm, by means of a 50 mm diam pipe 150 m long. Find the load that can be lifted at the crane hook and the speed of lifting if the velocity ratio of crane hook to ram is 6 to 1, the mechanical friction losses equivalent to 350 kN/m² pressure at the ram and the velocity of water in the pipe is 0·9 m/s. Take f = 0·01. State also the overall efficiency of the system.
Answer 1·12 metric tons, 0·1 m/s, 0·6 m/s, 0·905

2 The ram of a hydraulic press is 200 mm diam and is worked by an intensifier of the piston and the ram type which receives its low-pressure supply of water from a tank, whose surface level is 15 m above the intensifier piston, through a 50 mm diam pipe 120 m long. The intensifier ram is 75 mm diam and the piston is 900 mm diam. The mechanical friction of the packings may be taken as 3 per cent of the total pressure on the appropriate piston or ram. Calculate (a) the speed of ascent of the press when exerting a force of 500 kN, (b) the efficiency of transmission of the pipe, and (c) the overall efficiency of the machine. Take f = 0·0075.
Answer 24 mm/min, 79 per cent, 74·8 per cent

3 A hydraulic crane raises 2500 kg through a height of 14·4 m at a steady rate of 1·8 m/s once every 3 min. The pulley system attached to the crane has a velocity ratio of 8 and a mechanical efficiency of 0·75. The ram diam is 250 mm. The friction of the ram packing is equivalent to 210 kN/m² of ram area. Find (a) the pressure required in kN/m² in the cylinder to raise the load, (b) the pressure required to lower the load steadily, (c) the minimum

capacity of accumulator required in dm³ if the pumps work at a uniform rate.
Answer   5539 kN/m², 2788 kN/m², 84·4 dm³

4   A hydraulic crane having a velocity ratio of 6 is supplied from an accumulator in which the pressure is 5200 kN/m² by means of two pipes in parallel each 150 m long and for which the frictional coefficient is 0·01. The crane ram diameter is 175 mm and the efficiency of the pulley system is 70 per cent. Friction at the ram packing is equivalent to a pressure of 210 kN/m². The diameter of one of the supply pipes is 38 mm. Find the diam of the second pipe if the load which has a mass of 1400 kg is to be raised 0·9 m/s.
Answer   39·6 mm

5   The ratio velocity of the hook to velocity of the ram of a hydraulic crane is 8. The ram is 200 mm in diam and is fed from a hydraulic intensifier with diameters 150 mm and 600 mm. The low-pressure supply is at 170 kN/m². If the overall efficiency of the intensifier and crane is 70 per cent, find the speed of lift and the load carried by the hook when the system is utilizing 0·015m³/s from the supply.
Answer   760 kg, 0·24 m/s

6   The diameter of the ram of a hydraulic crane is 150 mm and the velocity ratio of the lifting gear is 6. The crane is supplied with high-pressure water from a main in which the pressure is 5500 kN/m² through a pipe 135 m long. Find the diameter of the supply pipe to the crane for a lifting speed of 0·75 m/s when the crane raises 1000 kg. Assume that the efficiency of the lifting gear is 66 per cent, that 5 per cent of the pressure operating the ram is lost in overcoming friction in the stuffing box, and that the friction coefficient f for the supply pipe is 0·01.
Answer   40·9 mm

7   A pair of railway buffers are used to arrest the motion of railway trucks. Each buffer consists of a piston working in a cylinder of 200 mm diam. The working fluid is water which is discharged to atmosphere through an orifice in each cylinder of 38 mm diam when the piston is forced into the cylinder.

If a truck having a mass of 2500 kg strikes the buffers when travelling at 32 km/h, determine the distance which each piston moves in reducing the truck speed to 8 km/h. Assume the coefficient of discharge of the orifice to be 0·65 and neglect the friction of the piston in the cylinder.
Answer   0·6 m

8   A pump supplies water at a steady rate to an accumulator which feeds a hydraulic crane working at a pressure of 4200 kN/m². The crane has to lift a body of mass 1500 kg at a rate of 0·9 m/s through 15 m once every 1½ min with an eficiency of 80 per cent. Calculate the volume of the crane cylinder, the minimum pump output power and the minimum capacity of the accumulator.
Answer   52·5 dm³, 3·07 kW, 42·7 dm³

9 A hydraulic crane has a jigger (a device for amplifying the movement of the ram), giving six fold multiplication. The ram is 200 mm diam and the cylinder is supplied with water through a pipe 300 m long. The pressure at the far end of the pipe is 8275 kN/m² and f (friction coefficient) for the pipe is 0·008. Find the diameter of the pipe so that a load of 2250 kg may be lifted at 6 m/s. Assume mechanical efficiency of crane to be 89 per cent. If a 75 mm diam pipe were provided, what would be the mass lifted and speed corresponding to the maximum power obtainable from the system?
Answer   73·6 mm, 2621 kg, 5·54 m/s

10 A hydraulic crane is to lift 2300 kg through 13·5 m at 1·35 m/s once every 2 min. The velocity ratio of the pulley system is 6 and its efficiency is 0·75. The crane cylinder is of 225 mm diam and the efficiency is 0·95 allowing for friction of the packing. The crane is fed from an accumulator, which is supplied at a steady rate from a pump. Calculate (*a*) the necessary pressure in the crane cylinder during lifting, (*b*) the duty of the pump in dm³/s, (*c*) the minimum capacity of the accumulator.
Answer   3770 kN/m², 0·746 dm³/s, 0·082 m³

11 A hydraulic machine delivering 33·5 kW with an overall efficiency of 75 per cent is operated by water supplied from a hydraulic accumulator through a 100 mm diam pipe, 90 m long. The accumulator has a 300 mm diam ram with a stroke of 3 m and carries a load which has a mass of 20 metric tons. The water for the system is supplied to the accumulator through a short pipe by a two-cylinder single acting pump having plungers of 200 mm diam, a stroke of 0·3 m and rotational speed of 40 rev/min.

Assuming 5 per cent slip in the pump and using a friction coefficient of $f = 0·007$ for the 100 mm diam pipe, calculate the longest continuous period during which the machine can be operated at full power.
Answer   47·7 s

12   The ram for a hydraulic crane has a diam of 150 mm and the ratio between the movement of the load and the ram is 6 to 1. Water is supplied through a 40 mm diam pipe having a length of 490 m, the pressure at the inlet end of the pipe being 7600 kN/m². The coefficient of friction for the pipe is 0·01. A pressure of 580 kN/m² on the ram is required to overcome mechanical losses. Determine (*a*) the maximum speed with which a mass of 1100 kg can be lifted, and (*b*) the load and speed of lifting which correspond to the maximum obtainable from the crane.
Answer   1·58 m/s, 1347 kg, 1·37 m/s

Index